INNOVATION AND VALUES

WILLIAM A. DOUGLASS
DISTINGUISHED SCHOLAR SERIES
NO. 3

Innovation and Values
A European Perspective

Javier Echeverría

Translated by
Cameron J. Watson

Center for Basque Studies
University of Nevada, Reno

This book was published with generous financial assistance from the Basque government

William A. Douglass Distinguished Scholar Series, No. 3
Series Editor: Sandra Ott

Center for Basque Studies
University of Nevada, Reno
Reno, Nevada 89557
http://basque.unr.edu

Book and cover design: Daniel Montero

Library of Congress Cataloging-in-Publication Data

Echeverría, Javier.
Innovation and values : a European perspective / Javier Echeverría ; translated by Cameron J. Watson. -- First [edition].
pages cm. -- (Douglass scholar series ; No. 3)
ISBN 978-1-935709-55-8 (pbk.)
1. Technological innovations--Economic aspects. 2. Technological innovations--Philosophy. 3. Research, Industrial. 4. Technological innovations--Government policy--Europe. 5. Research, Industrial--Goverment policy--Europe. I. Title.

HC79.T4E225 2014
338'.064094--dc23

2014032331

CONTENTS

PROLOGUE

In 2000, the European Union (EU) implemented the Lisbon Agenda in order to stimulate the knowledge economy in Europe. It set forth an ambitious objective: that the EU would, by 2010, be the world leader of the knowledge society, overtaking the United States and Japan in this area of the economy. Its key instrument would be RDI (research, development, and innovation) policies, both within the European states and the Union itself. On the basis of this strategy, the Sixth Framework Programme was designed (2002–2006), followed by the Seventh (2007–2013), which came to an end while this book was being written.

In 2005, the European Commission (EC) evaluated the Lisbon Agenda and confirmed that the proposed objective would not be achieved by 2010, or in the years that followed. The EU had attained positive results in R&D, but not so much in innovation. The "European Paradox" had once again been repeated:[1] Europe produces quality scientific knowledge and technologies, but this is not translated into a noticeable growth in its innovative capacities. In 2005, the United States and Japan continued to enjoy a clear advantage over Europe in this aspect, and globally, new competitors or emerging countries (the BRICS, or Brazil, Russia, India, China, and South Africa) had emerged. Their scientific level was below

1. European Commission, *Green Paper on Innovation*, http://europa.eu/documents/comm/green_papers/pdf/com95_688_en.pdf.

that of the EU yet their economies were extremely competitive in global markets.

Other factors were influential in the relative failure of the Lisbon 2000 European Agenda . One important factor, which I will not discuss in this work, was the rejection of a projected single European constitution by some countries. This document was agreed on by the EU heads of state and governments on October 29, 2004 in Rome. However, it had to be ratified country by country, and it was soon clear that there would be no unanimity among the European states. Therefore the European Council, meeting on June 16 and 17, 2005, decided to open a period of reflection on Europe's future. One of the questions to be considered was the Lisbon Agenda, which was subsequently modified (the Renewed Social Agenda, July 2, 2008) and then redefined by means of a new strategy, Europe 2020.

Europe 2020 included seven emblematic initiatives, one of which, termed Innovation Union[2], referred to European innovation policies. Innovation Union is presently underway and will take more shape in the Horizon 2020 program, which will replace the Seventh Framework Programme from January 2014 until the end of 2020. Conceivably, Horizon 2020 will bring with it profound changes and innovation polices, precisely at a time when Europe has been badly affected by the financial crisis, which began in 2007 and brought a severe economic depression, especially to some countries in Southern Europe (Greece, Portugal, Spain, Italy, and Cyprus). Innovation is required to escape the crisis, yet the models that have shaped European innovation policies are being called into question, in particular that of RDI. There have been mistakes in European polices, but the main cause of failure as regards the 2000 Lisbon Agenda is due not just to political errors but also to the profound change experienced by the concept of innovation throughout the first decade of the twenty-first century. The OECD (Organisation for Economic Co-operation and Development), one of the main global reference points in the study of innovation, makes it clear in its report *The New Nature of Innovation* (2009) that this transformation has already taken place and that it

2. European Commission, *Europe 2020*, 546 final.

is necessary to reconceptualize innovation. The same is true of the EU, although more recently, and still without any clear consensus.

This strategic shift is taking place because in recent years new types of innovation, aside from technological innovation based on R&D, have emerged. For a country, region, institution, or company to be innovative, R&D is no longer enough. Scientific knowledge is still important for some (the most disruptive) innovations, but other innovation processes require different abilities and skills as well as other categories of knowledge. In its Lisbon Agenda, the EU backed the economy and the knowledge society. Ten years later, it preferred to prioritize the idea of an innovating Europe.[3] These are not incompatible objectives, but the order of priorities changed. In 2000 the priority was knowledge; in 2010 it was innovation, even though the quest for knowledge continued to be a significant goal.

This shift toward innovation has been evident not just in policies but also in studies of innovation, which have proliferated in the last five years. This will be the main topic of the present book. I will start by discussing the linear model, according to which companies undertake innovation as a result of their own R&D knowledge and that transferred to them by universities and research centers. Vannevar Bush proposed the linear model in his celebrated report, *Science, the Endless Frontier* (1945), to President Roosevelt. Its application was decisive in science policies centered on "Big Science."[4] In the early 1980s, a new category of science, technoscience, emerged whose principal feature was that knowledge played a subordinate role to innovation. Silicon Valley was a good illustration of this profound transformation, together with the emergence of digital information and communication technologies, which radically transformed the sciences, engineering, and the arts, as well as businesses, organizations, and societies. The technoscientific revolution was based on the linear model and the ideas of Joseph Schumpeter, according to which innovation is the driving force of economic growth.

Since then, new forms of innovation have emerged, apart from

3. Aho et al., *Report of the Independent Expert Group on R+D and Innovation*.
4. Weinberg, "Impact of Large-scale Science on the United States"; Solla Price, *Little Science, Big Science*.

products created by the manufacturing industries: for example, innovation in processes, innovation in service sectors, organizational innovation, and marketing innovation.[5] The emergence of information and knowledge economies and societies has brought with it a proliferation of innovation methods, and with that the inadequacy of the linear model. It has been noted on repeated occasions that the linear model has been overtaken, yet this is questionable, at least in Europe. To my mind, the subordination of innovation policies to R&D policies has been one of the main causes of the 2000 Lisbon Agenda's relative failure, even though the linear model may have been replaced by the systemic model.[6] Nor should one forget that this latter model has also been called into question: "the innovation system concept is typically a European concept that is not always used so explicitly in non-European countries."[7]

In sum, there are particularly important conceptual, epistemological, and methodological problems when it comes to analyzing innovation processes. No matter how much there has been talk in Europe of local, regional, and national innovation systems, these systems have continued to employ R&D as the basis of innovation policies. This is the main reason for the European paradox, as I will attempt to demonstrate in this book and especially in chapter 4. Prior to this, I will devote some time to the origins and development of innovation studies, which emerged at the same time as technoscience during the final decades of the twentieth century. I will likewise argue that the dominant paradigm in innovation studies, the OECD's *Oslo Manual* (1992, 1997, and 2005), has also entered into a period of crisis. That crisis is demonstrated through multiple anomalies in the paradigm and above all by the emergence of an alternative paradigm, that of social innovation (chapter 3).

In order to understand this transformation one must address the history of innovation. In chapter 1, I will evoke Schumpeter, the first great theoretician of innovation, underscoring his distinction between invention and innovation, and his conception of in-

5. See OECD and Eurostat, *Oslo Manual.*

6. Lundvall, ed., *National Systems of Innovation*; Nelson, ed., *National Innovation Systems.*

7. Boekholt, "The Evolution of Innovation Paradigms," 345.

novation as the creative destruction typical of capitalism. I will also distinguish between science (aimed at the quest for knowledge) and technoscience (aimed at achieving innovations), comparing that distinction to other theoretical alternatives (Mode 2 knowledge, postacademic science, postnormal science, and the actor-network theory) that have been proposed as a means of interpreting the major transformation in science at the close of the twentieth century. Personally, I prefer to speak about the late twentieth-century technoscientific revolution, out of which innovation studies emerged.

Chapter 2 will examine in detail the linear model and its transformation into a systemic model.[8] The *Oslo Manual*, which has changed significantly through its three editions (1992, 1997, and 2005), will be the main point of reference in the first part of the book, in which I will discuss the first signs of crisis in the linear model, at least in studies of innovation. From 2005 onward, and even somewhat beforehand, there was the first inkling of an alternative paradigm to technological innovation: namely, social innovation. I will explore the emergence and detail of social innovation in Canada and Europe in some detail in chapter 3. Chapter 4 analyzes the crisis of the dominant paradigm in depth, using as its starting point the concept of hidden innovation as proposed in 2006 and 2007 by the National Endowment for Science, Technology and Arts (NESTA, now renamed Nesta) in Great Britain. In order to study the different categories of hidden innovation (in the public sector, social innovation, user innovation, innovations in the cultural and artistic sectors, and so on), one must set out from a new conceptual and methodological framework, as I will attempt to demonstrate in chapter 4. There I will discuss in some detail the full gamut of innovation categories that are not based on R&D and that, as a result, call into question the dominant paradigm. In 2009 and 2010, the OECD itself acknowledged that the conceptual framework for innovation studies would have to be radically modified, despite the fact that it has still not published a fourth edition of the *Oslo Manual*. In contrast, some European

8. Lundvall, ed., *National Systems of Innovation*; Nelson, ed., *National Innovation Systems*.

countries have developed instruments to measure non-R&D innovations, such as the Measuring Public Innovation in the Nordic Countries (MEPIN) project, the Innometrics customer data management platform, and the Innobarometer survey of the European Commission. And they have been joined in these endeavors by global institutions such as the United Nations (UN)/ United Nations Conference on Trade and Development (UNCTAD) and the United Nations Educational, Scientific and Cultural Organization (UNESCO), which have changed the methods by which they measure culture and creativity.

All these debates indicate the need to think quite deeply about the concept of innovation, which in turn leads to wide-ranging ontological, methodological, epistemological, and axiological problems. The category of process is difficult to analyze: there is no commonly established definition of innovation but several kinds, as well as scales, of innovation; there are not just agents who promote innovation but also obstacles and agents interested in its effects and consequences (stakeholders); it is not at all easy to measure innovation; and, last but not least, innovation processes also occur in natural environments, not just in human societies or markets. All these problems are mentioned and analyzed briefly in chapter 5, in which I contribute my own proposals, made from a philosophical perspective, to innovation studies. There I propose a *naturalized innovation* and insist on the importance of the axiological dimension to innovation processes. Generating value is the main feature of innovation, apart from its procedural condition. However, I am not just referring to economic value, like that produced in the markets, but also to other types of value: social, epistemic, political, environmental, and so on. Ultimately, I suggest a philosophical definition of innovation that could be useful, in my view, because it embraces most of the definitions used by experts in the field: *innovations are interactive processes that generate something new and valuable in specific environments*. It is enough to distinguish different types of value so that the different categories of innovation might easily fit that definition.

Innovation studies must be interdisciplinary, given the difficult nature of their object of study. Until now, economists have dealt with the subject in the wake of Schumpeter and his followers, the evolutionary economists. The very expansion of innovation in different countries and sectors has given rise to a real proliferation of meanings that threaten to convert the concept of innovation into something analytically worthless. In order to reduce this risk, one must have at one's disposal definitions and instruments of analysis and measurement that can be applied to the distinct types and agents, as well as to the diverse contexts in which innovation processes are developed.

Chapter 5 of this work seeks to make an initial philosophical contribution to those studies that offer a minimum degree of rigor and accuracy, helping to elucidate the concept in question instead of just complicating its use further as many of the current ideologues and publicists of innovation do. I hope that this book, as well as contributing data and references about the development of innovation policies and studies in Europe, marks the beginning of a new line of historical and philosophical research that takes into account the history and philosophy of science and technology, but which underscores the profound specificities of the concept of innovation. It was, originally, linked to science and technology. Nowadays, innovation processes have been developed and diversified to such an extent that the linear model is clearly inadequate, if not erroneous, at least when it comes to shaping innovation policies in Europe.

Acknowledgements

I began to prepare this book in Reno, Nevada, where I was a visiting researcher at the Center for Basque Studies (CBS), University of Nevada, Reno, in my capacity as Douglass Visiting Scholar. Spending just under one year there (September 2008 to July 2009) was particularly enjoyable for my wife, daughter, and I thanks to the care, kindness, and warmth of several people at the CBS: Joseba Zulaika (its then director), Sandy Ott (faculty member), Bill Douglass and Carmelo Urza (emeritus faculty), and Pedro Oiarzabal (researcher), together with Kate Camino, the CBS office manager, who greatly facilitated the different administrative formalities that I had to fulfill during that academic year. I extend my thanks and friendship to all of them.

The text in this book revisits and restructures certain articles and book chapters that I previously published on issues in technoscience and innovation, all of which I cite in the bibliography. I am grateful to these publishers for allowing me permission to make use of these materials. Likewise, this book addresses certain issues associated with the INNOC (Innovaciones ocultas: cambio de paradigma en los estudios de innovación, Hidden Innovations: The Changing Paradigm in Innovation Studies) research project, which was financed between January 2012 and December 2014 by the Spanish Ministry of the Economy and Competitiveness (project FFI2011-25475). I am carrying out the INNOC project at the Department of Sociology 2, University of the Basque Country, as a Research Professor employed by the Basque Science Foundation, Ikerbasque. My thanks to Mari Carmen Gallastegui, Ander Gurrutxaga, Alfonso Unceta, and Fernando Cossío for their academic and institutional support during this latter period of compiling the book. Its publication was delayed more than anticipated as a result of the important changes experienced by innovation studies in Europe during this time. That forced me to rewrite, on more than one occasion, certain sections in the five chapters that were planned from the outset.

Both in Reno and Donostia-San Sebastián, it would have been impossible to write this book without the strong emotional and affective support of my wife, Belén Altuna, and my daughter, Irene Echeverría Altuna. I dedicate this work to both of them.

—Javier Echeverría
December 31, 2013

1

SCIENCE, TECHNOLOGY, AND INNOVATION IN THE TWENTIETH CENTURY

Evoking Schumpeter: Innovation as Creative Destruction

Joseph Schumpeter is usually taken to be the first great theoretician of innovation, although he was influenced by authors such as Karl Marx and Werner Sombart in the sense that they, too, pointed out the dual condition, both destructive and creative, of capitalism.[1] Schumpeter also took on board Marx's thesis according to which capitalism never arrives at a stable equilibrium but rather, because of its very structure, tends to create discontinuities and crises, destroying previous forms of wealth and creating others. However, his explanation for these imbalances is radically different to that of Marx. According to Schumpeter, innovation is the driving force of capitalism and the particular cause of its cycles, even when there may be other structural factors to bear in mind, such as natural resources, the previous state of the economic system, capital, technologies, work, and interests. Using the primacy of innovation as a starting point, a topic that already appears in his youthful writings as well as in his doctoral dissertation,[2] throughout his life the

1. See Reinert and Reinert, "Creative Destruction in Economics," 55–86. This work highlights the influence of Nietzsche's *Thus Spoke Zarathustra* and notion of the Superman on Schumpeter and his description of the innovative entrepreneur. Likewise, it points out that the German economist Sombart used the expression "creative destruction" before Schumpeter (56).
2. Published as Schumpeter, *Theorie der wissenschaftlichen Entwiklung*;

Austrian economist proposed a theory that contributed the first conceptual framework to innovation studies. Indeed, Schumpeter's ideas continue to exert a strong influence in the currently dominant paradigm in the field: the OECD's (Organisation for Economic Co-operation and Development) *Oslo Manual* and Eurostat.[3] Meanwhile, his ideas served as an inspiration for authors like Richard R. Nelson and Sidney G. Winter who, with their publication of the article "In Search of a Useful Theory of Innovation" in 1977, encouraged the notion of evolutionary economics. Given the fact that, in this work, I will propose an evolutionary conception of innovation and an alternative paradigm to that of the *Oslo Manual*, I will first analyze these two lines of thought.

Schumpeter reflected deeply on the future of capitalism, ultimately arriving at a prediction that it would collapse and that it would be replaced by some form of socialism. Marx had prophesized something similar, which is why Schumpeter was very careful to distinguish his economic theories from, and pit them against, those of Marx. In order to do so, he postulated the existence of a new source of creating wealth, namely that of innovation rather than through the force of labor. According to the Austrian economist, innovative entrepreneurs are the driving force of capitalism, but with one important caveat: they do not just create wealth but also destroy it. Thus, innovation implies a *creative destruction* that characterizes capitalism: indeed for Schumpeter, "This process of Creative Destruction is the essential fact about capitalism."[4]

He dedicates the first part of his book *Capitalism, Socialism and Democracy* (1942) to analyzing and critiquing Marx's theories. Despite his radical differences with Marxism, he constantly praises some of Marx's contributions, in particular the idea that the very development of capitalism implies significant crises and even the possibility of its eventual downfall. In the second part of the book, titled "Can Capitalism Survive?," Schumpeter outlines and makes the case for his own ideas in this regard. Among his many propos-

English trans., *The Theory of Economic Development.*

3. OECD and Eurostat, *Oslo Manual.* Eurostat is part of the European Commission. Its task is to draw up and provide statistical information to EU institutions.

4. Schumpeter, *Capitalism, Socialism and Democracy*, 83.

als, I will mention seven that, to my mind, are the most significant when it comes to innovation studies:[5]

1. The capitalist economy is not, nor can it ever be, stationary; it is always in continuous change and evolution.[6]
2. The driving force of such change is innovation.[7] The nucleus of his theory about capitalism and his main argument against Marx is rooted in this point.
3. One must differentiate between invention, innovation, and diffusion. Inventors contribute new ideas, but innovators introduce them in practice; they undertake them.[8] Later, other competitors imitate or emulate these innovations, thereby contributing to diffusing them throughout entire economic sectors. Advertising and patent sales are typical procedures used to diffuse such innovations. The success or failure of an innovation almost always depends on this third phase of diffusion and commercialization.
4. Entrepreneurs encourage innovation. This thesis is key to understanding the thought of Schumpeter, who made a great effort to define the figure of the entrepreneur and distinguish entrepreneurs from the investors, administrators, and managers in a company. He defines entrepreneurs according to their function in an economic system: "the function of entrepreneurs is to reform or revolutionize the pattern of production by exploiting an invention or, more generally, an untried technological possibility for

5. Ibid., 31–32.

6. "Capitalism, then, is by nature a form or method of economic change and not only never is but never can be stationary." Ibid., 82.

7. "The fundamental impulse that sets and keeps the capitalist engine in motion comes from the new consumers, goods, the new methods of production or transportation, the new markets, the new forms of industrial organization that capitalist enterprise creates." Ibid., 83.

8. The function of the entrepreneur "does not essentially consist in either inventing anything or otherwise creating the conditions which the enterprise exploits; it consists in getting things done." Ibid:, 132. In another work, Schumpeter is even more explicit: "Innovation is possible without anything we should identify as invention and invention does not necessarily induce innovation, but produces of itself no economically relevant effect at all." See *Business Cycles*, 80.

producing a new commodity or producing an old one in a new way, by opening up a new source of materials or a new outlet for products, by reorganizing an industry and so on."[9] Later I will go back to this functional notion of entrepreneurship and extend it to innovation processes.

5. Schumpeter distinguishes five types of innovation: product, process, organizational, the discovery of new raw materials, and the opening of new markets. With some variations, all five continue to be important for what I would term the "Oslo model" that currently dominates innovation studies. Thus, Schumpeter's argument has been a major reference point. He focused on disruptive innovations, those that can end up transforming the productive system or at least part of it, in contrast to cumulative innovations, which only contribute minor improvements. Creative destruction emerges out of the former, to the extent that they radically transform the productivity or competitiveness of businesses and, in the final analysis, the structure of markets. The distinction between disruptive and cumulative innovations has become a standard reference in innovation theories. Elsewhere, Schumpeter argues that entrepreneurs do not form a social class, but can in fact emerge out of any social stratum.[10] Put in current terms: he was interested in entrepreneurs and entrepreneurship. He also attributed leadership qualities to entrepreneurs, but without identifying either figure specifically.
6. Innovation is essential to the capitalist economy because it reactivates the productive forces every time it transforms the systems of production. However, major innovations also cause crises, which is a point Schumpeter always emphasized. When a disruptive innovation occurs, many companies do not adapt to the new situation and disappear, with the resulting loss of wealth and jobs. Schumpeter even argued that, although there was no doubt that

9. Schumpeter, *Capitalism, Socialism and Democracy*, 132.
10. For Schumpeter, "though entrepreneurs do not per se form a social class, the bourgeois class absorbs them and their families and connections." Schumpeter, *Capitalism, Socialism and Democracy*, 134.

major innovations created prosperity, they could also be the cause of economic recessions following a cycle of prosperity.[11] Such hypotheses are very revealing when it comes to interpreting a crisis like the current one (2008–2012), whose origins lie in a series of financial innovations but that later affected the global finance markets and later whole countries.

7. In capitalist economies innovations imply creative destruction. I will return to this particular Schumpeterian thesis throughout the work. For the time being, I will limit myself to citing a passage by Schumpeter in which he contends that creative destruction might ultimately jeopardize the capitalist system itself: "the capitalist process in much the same way in which it determines the institutional framework of feudal society also undermines its own."[12] Indeed, in *Capitalism, Socialism and Democracy* he leaves no doubt as to the destructive condition of capitalism and analyzes it at length.[13]

The seven proposals listed here deserve a more detailed discussion. I will focus on points 3, 4, 5, and 7, which are the most important as regards innovation studies. The seventh point, specifically, is one of the most controversial and is philosophically very substantial.

I will then advance some initial ideas in this regard. Many late twentieth- and early twenty-first-century practitioners of innovation, as well as some policymakers, typically formulate beatific and sanctimonious discourses about innovation in which they only highlight its positive aspects. European Union documents abound in which it is argued that innovation is a fundamental part of the renewed Lisbon Agenda (*Union Innovation 2020*) because it can help increase the competitiveness of European businesses and contribute to economic growth, social well-being, and job creation. Schumpeter, one of the major twentieth-century apologists of capitalism, was at least honest enough to state clearly that, in a capi-

11. See Schumpeter, *Business Cycles*, 142–50.
12. Schumpeter, *Capitalism, Socialism and Democracy*, 139.
13. See, for example, ibid., 135.

talist economy, innovation produces both gains and losses. Such considerations tend to be absent from documents drawn up by European policymakers, which is precisely why it is so necessary to invoke Schumpeter. He was quite clear about the fact that innovation was both creative and destructive and he referred above all to entrepreneurial innovation. Later, I will contend that innovation is not just carried out by entrepreneurs. As a result, I will critique a fundamental point of his theory. Nevertheless, other contributions of his are still valid and even applicable to other forms of innovation, including social, cultural, and artistic. Because Schumpeter was an economist, it makes sense that his analysis was restricted to entrepreneurial innovation. For my part, I attempt to explore the creative destruction hypothesis in other kinds of innovation, including innovations in the public sector and innovations by users.

Once this first working hypothesis, which will be applied in the following pages, has been formulated, it will be expedient to extract some of its methodological consequences. Accepting the creative destruction hypothesis in innovation studies implies, at the very least, four methodological rules:

1. In every empirical study undertaken, one must seek out both the benefits and harm produced by innovation processes.
2. One must detect, identify, and analyze the potential obstacles to innovation because they also form part of the value chain.[14] In accordance with this methodological rule, it is not just innovative agents (entrepreneurs, facilitators) that are of interest but also resistance and the obstacles to innovation (delaying and hindering factors). Because innovations are destructive, logically there always will be social agents who are threatened by and who will resist innovation.
3. One must analyze innovation processes on the basis of the consequences derived from them. This methodological rule makes it advisable to evaluate not just the immediate results of an innovation process but also its final conse-

14. The value chain idea is central to Michael E. Porter's model in *Competitive Advantage*.

quences, some of which might be sustained by agents who were not initially involved in the innovation process: for example, workers who were made redundant as the result of a recession created by disruptive innovations.

4. One must pay attention to innovation processes that fail instead of just concentrating on successful innovations. It is typically estimated that most innovations, however promising they may appear initially, are unsuccessful when it comes to their process of diffusion or commercialization. From the perspective of innovation studies, such case studies can be equally or even more illustrative than their successful counterparts, because they allow one to identify different factors that form part of an innovation process.

Another Schumpeterian proposal that I find completely acceptable consists of the distinction between inventors and innovators. The former are characterized by their creativity and the latter by their entrepreneurial ability. Inventors contribute new ideas, while innovators introduce them in practice. Naturally, through history there have been inventors who, as well as conceiving, have also developed their inventions and attempted to apply them to resolving specific problems. Archimedes, Leonardo da Vinci, and Charles Babbage are three examples of this kind of inventor, among many others that could be mentioned. In differentiating innovators from inventors, Schumpeter does not indicate two separate classes but rather argues in favor of something more important as regards innovation studies: thinking something out and invention are two distinct phases in the process, but there are others, such as diffusion and, where appropriate, commercialization. In order to be an innovator it is not enough to be an inventor, although this does not lessen the interest of inventors when it comes to drawing up a theory of innovation because they do play an important role in the value chain. Entrepreneurs must be attentive to new ideas, but their principal task is to promote them, develop them, and add value to them. In sum: economic innovations are produced in markets, not in laboratories, workshops, or offices. New ideas and inventions are the first links in an innovation process, but other agents intervene throughout this process: competing companies, clients, regulators, and so on. In distinguishing between invention, innovation, and diffusion, Schumpeter prefigures the linear model of innovation. This model, which is usually rendered by the letters RDI, dominated twentieth-century innovation studies, as we will see in the next chapter.

A Schumpeterian Definition of Innovation

How does Schumpeter define innovation? In order to answer this question one would have to analyze his work in detail, which is not the goal of the present work. I will just discuss some of the definitions he offers, especially those based on the notion of *new actions*. According to him, when there is an important change in an economic system, whether because of a significant increase in population, because new competitors emerge, because working conditions change, or for whatever other reason, there are two kinds of potential responses: adaptive and creative.[15] Through the former, businesses redesign their objectives and strategies, trying to adapt them to the new conditions of production and the market. He characterizes the latter according to three distinctive features:[16]

1. Creative responses can be explained *ex post*, but not *ex ante*. Put another way, they cannot be predicted and may, therefore, catch competitors by surprise, in contrast to adaptive responses, which are predictable.
2. Creative response shapes the whole course of subsequent events. In other words, it transforms markets and situations, sometimes radically. For Schumpeter, creativity is never innocuous; it is always transformative, and even destructive. Obviously, he means disruptive innovations.
3. Creative response has something to do with the personnel available in a specific social context and, more specifically, with certain individuals – namely, those people who encourage such creative responses. Schumpeter classifies this kind of person as an entrepreneur.

15. Schumpeter, "The Creative Response in the Economic History."
16. Ibid., 150.

For my part, I will concern myself with the emergence of creative responses in other forms of innovation, especially social, cultural, and artistic innovations.

According to Schumpeter, innovations are produced in different kinds of societies,[17] but they are characteristic of competitive free-market economies in which innovative entrepreneurs are the driving force of economic growth. He then goes on to state that, "Seen in this light, the entrepreneur and his function are not difficult to conceptualize: the defining characteristic is simply the doing of new things or the doing of things that are already being done in a new way (innovation)."[18] I would call the attention of readers to these words because they permit me to take my first step in this inquiry. Schumpeter termed innovators entrepreneurs and, because he was an economist, was mostly interested in innovations stemming from businesses. However, this quote contains a characterization of the function of entrepreneurs that goes beyond that of the economy and that can be applied to other areas of human action: *the doing of new things or the doing of things that are already being done in a new way*. Inventors usually have new ideas, but innovators go a step further: they move on to act and make these ideas a reality, at least in part. Moreover, Schumpeter points out that he is contributing a defining characteristic of innovation; this is not just some passing comment.

This initial foray into his ideas allows me to come up with a "Schumpeterian definition" of the concept of innovation: *innovation consists of the doing of new things or the doing of things that are already being done in a new way*.

In what follows, I will demonstrate that this definition can be improved on, and from several perspectives. Nevertheless, it opens up a means of reflection that is worth pursuing because it links the theory of innovation to the theory of action. We thus have at our disposal an initial conceptual framework in order to inquire into the concept of innovation. To put it in the typical jargon: doing new things entails *product innovation*, while doing things that are already being done in a new way is equivalent to *process innovation*.

17. Ibid.
18. Ibid., 151.

Whatever the case, a theory of innovation for the twenty-first century must go beyond the economy and be based on a theory of new actions. Information and communications technologies (ICTs) are good illustrations of such action-based innovations.

Let us examine some of the weaker aspects of this initial Schumpeterian definition. Clearly, actions and innovative activities are what is important here, as well as the agents who encourage such activities and the agendas they use. In principle, these are entrepreneurs, yet I have already indicated that, even though entrepreneurs drive innovation processes, they need the help of other agents to make innovative ideas successful. For example, one should also study the reaction against or resistance to innovation. Those that criticize and oppose innovative ideas are also stakeholders in an innovation process.

A second problem derives from the use of the term "things," which is excessively ambiguous. Economists today prefer to speak about new products and new processes instead of new things. In any event, allusion is being made here to the results of certain innovative actions.

From a philosophical perspective, one should initially define innovation as the emergence of new actions and new ways of doing something, independent of whether such actions are of an economic nature or not. Doing something new, insofar as a product that did not previously exist is created, implies innovative action independent of the specific product that results from such action.[19] In contrast, producing something that already exists, but in a new way, entails an innovation process. Extending Schumpeter's ideas, I would say that a philosophical notion of innovation must be based on two categories: that of *doing* and that of *ways of doing*. The advantage of this initial Schumpeterian-inspired definition of innovation is its general nature. It does not just allude to economic actions but to any kind of human action.

I will offer a very straightforward example. Let us consider a mathematical problem, such as squaring the circle. One of the

19. Scientific innovations are actions that produce new theories and/or new scientific facts. Disruptive innovations are those that call into question existing paradigms, in other words, which imply radical contributions to knowledge.

great seventeenth-century (and even earlier) mathematical contributions was to demonstrate that it is not possible to square a circle with ruler and compass. At the time, this was a great mathematical innovation because it contributed something radically new, precisely because this problem had been investigated since Ancient Greek times without finding any solution. One could say much the same for Kurt Gödel's theorem or Andrew Wiles's proving of Fermat's last theorem. Mathematical proof can be highly innovative when it manages to prove something that no other mathematician has managed to prove before. However, once something has been proven, managing to prove it by other means is also a significant mathematical contribution. Paraphrasing Schumpeter, it implies proving something that has already been proven, but doing so by means of a new and until this time unknown procedure. Although the Austrian economist was only thinking about the economy when he advanced the above-mentioned definition, his ideas can be applied to mathematics and many other sciences. For example, we can distinguish between an experiment that proves a new fact and an experiment that proves once again an already proven fact albeit in a different way (using other apparatuses, following other procedures, under more general conditions, and so on). The Schumpeterian definition of innovation (doing new things, doing them in a new way) is a good starting point for any philosophical reflection.

The Austrian economist himself discusses this initial definition of innovation in the following terms: "In fact it is an advantage, that such a definition does not draw a sharp line between what is and what is not 'enterprise.' For actual life itself knows no such sharp division."[20] The concept of the businessman (entrepreneur) can be extended, and with it the notion of innovation as well. This is one of the conceptual undertakings, forgive the repetition, that I seek to undertake in this book: namely, to set out from the rigid core of Schumpeterian thinking to extend the notion of innovation to other areas of human action. Seen from this perspective, Andrew Wiles undertook an important mathematical undertaking when he withdrew from public life in order to dedicate his time

20. Schumpeter, "The Creative Response in the Economic History," 151.

exclusively to proving Fermat's last theorem. In the end he was successful, but he had made a prior decision to become an entrepreneurial and innovative mathematician, instead of limiting himself to continue getting cited in high-impact journals. One might say the same for any scientist who designs and carries out a new experiment, since he or she might end up producing a new scientific fact. One example of this is the Michelson-Morley experiment on the deviation of light rays, which was the first verified empirical support for Einstein's theory of relativity. One could say the same for the Higgs boson at the CERN (European Organization for Nuclear Research) in Geneva. Whenever an individual scientist or a research team undertakes the task of doing a new experiment and discovering a new scientific fact that transforms previously accepted knowledge, *scientific innovation* has taken place. In this way, the notion of *scientific discovery* comes to be associated with that of innovation, which could be productive for the current study. In sum: both in the case of the formal sciences and in that of the physical and natural sciences, close links can be established between theories of *scientific change* and theories of scientific innovation.

What I have just suggested does not just concern *scientific doing* but also other kinds of human doing such as social, cultural, and artistic action as well as many other forms of human action. In principle, it seems possible to apply the different forms of human action to the wider notion of innovation that, reinterpreting Schumpeter, I am proposing here:

> Definition 1: *Innovation is doing something new, or doing something that has already been done, but in a different way.*

This definition will not be my definitive proposal for this book. Rather, I will offer that in chapter 5, but in order to get there one should first clarify the concept of innovation. In this initial definition, I link it to doing and to ways of doing, not just to the idea of something new. According to Schumpeter, there is no innovation without an innovating agent (normally an individual, the entrepreneur) who promotes and stimulates the innovation in question. For my part, I will call into question this entrepreneurial reductionism, because a number of different agents take part in the innovation process: no innovation is successful without a certain degree of social acceptance, whether in markets or in other scenarios (which also exist, given that it is not just all about markets). Having clarified this, I would say that definition 1 is a good foundation for my conceptual investigation, even though this definition will change quite a lot when, in chapter 5, the "final" definitions of innovation are presented. Examining what innovation might be is, after all, also an innovation process — in this case, a conceptual one. Schumpeter adds three further requisites necessary to define the concept of innovation to those features of creative responses cited at the beginning of this section:[21]

(a) The emergence of an innovation cannot be predicted from preexisting facts.
(b) An innovation changes the course of subsequent events, including their long-run outcome.
(c) Innovative actions stem from people and in particular from individuals.

Requisite (a) is very important in innovation studies because, if true, it would mean that such studies will never be predictive. As such, this leads to the problem of whether there can ever be innovation sciences or not, and if such sciences would only be interpretative and, at most, explicative, but never predicative. One might even speak in terms of innovation arts. This problem emerges because innovation processes take place in conditions of uncertainty, as repeatedly highlighted by Ander Gurrutxaga.[22] This is something I will have to bear in mind throughout. Requisite (b) offers a philosophically important consideration, for two reasons: First, because an innovation results in several consequences, both advantageous and disadvantageous, which lead at least to reevaluating, and possibly redirecting, the process of innovation itself; all this implies a consequentialist conception of innovation in which one takes into account not just immediate but also long-term results. Second, because some innovations do not just contribute change

21. Ibid., 150.
22. See, for example, Gurrutxaga, Pathways to Social Innovation in the Basque Country, 53.

in the field in which they take place but also radical changes more generally, in other words transformations. Schumpeterian innovation theory is linked to both theories of change and theories of transformation in the environment in which the innovation occurs. In fact, Schumpeter always contended that authentic innovation implied considerable changes in markets and therefore creative destruction. The destructive condition of Schumpeterian innovations allows one to speak about new actions and new ways of doing that transform the economic world, for example by creating new products and processes that did not previously exist. This requisite could be incorporated into a definition of innovation, in which case one would be speaking not just about new actions, but about new transformative actions in the field in which they take place. As regards requisite (c), it raises one of the main issues for innovation studies: on the premise that innovations refer to new actions, who are the agents that promote and develop these new actions or new ways of doing something? The problem of innovative actions leads to the question of innovative agents, whether these are individuals (as Schumpeter argues) or collective agents such as organizations, cities, regions, countries, or societies. Schumpeter was a committed methodological individualist, and not for nothing did he invent the expression "methodological individualism," which years later turned into an important conceptual innovation among philosophers of science. However, should methodological individualism be one of the basic principles in innovation studies, whether inspired or not by Schumpeter? For my part, I believe one should also take account of collective innovative agents, for example when group work is undertaken or that of institutions or innovative companies. I will opt for a pluralist solution: there are individual, collective, and institutional innovative agents, but also barriers to innovation that are individual, collective, and institutional. In order to analyze and measure innovation processes, one must take all stakeholders into account, not just the self-made men who underpin the romantic myth of the Schumpeterian individual entrepreneur, as well as self-made women, whom Schumpeter does not mention!

Qualifying this third requisite in the form of a fourth, Schumpeter adds an important comment:

> (d) "It should be observed at once that the 'new thing' need not be spectacular or of historic importance. It need not be Bessemer steel or the explosion motor. It can be the Deerfoot sausage."[23]

In other words, innovations can be produced at the micro-, meso-, or macroscale. The Internet is a good example of macro-innovation in that its diffusion has transformed the economies of many countries and regions, as well as the lives of millions of people. In contrast, the construction of the Guggenheim Museum Bilbao represented a mesoscale innovation in that it only transformed the city of Bilbao and its immediate surroundings. There are innovations that only transform microcosms, but the transformations they induce in them can be radical: renovation work in a house, for example, or moving into a new apartment, neighborhood, or city. For those that experience such changes, their *Lebenswelt* (Life world) is radically transformed: these are examples of new spaces of action, although taking place on a small scale. The innovation studies I advocate in this book must address diverse kinds of innovation (technological, business, social, artistic, and so on) but also at different levels in innovation processes. Some transformations only affect human microcosms. However, if they change them completely, they are innovations in the true sense of the word even though they only take place at the small scale.

Taken together, as regards this "first definition," Schumpeter's four requisites hint at just as many additions and improvements. We are not just interested in new ways of doing and new products; in order to be considered authentic innovations, such actions and procedures must entail significant transformative consequences, regardless of whether these are produced at the micro-, meso-, or macroscale. Thus emerges a "second definition" of innovation based on Schumpeter's ideas:

> Definition 2: *Innovations are new actions (and ways of doing) that lead to significant transformations in something (let us*

23. Schumpeter, "The Creative Response in the Economic History," 151.

term it X), whether at the micro-, meso-, or macroscale.

It remains to be explored what is actually transformed by innovative actions, which is something I will return to later. In Schumpeter's case, economic markets are transformed, whether through new products, new processes, new forms of organization, new markets, or new primary materials. A disruptive innovation must have some visible effects in some economic sector. These are positive for some and negative for others because it implies creative destruction. For that reason, the effects of the innovation must be valued both positively and negatively. Let us briefly analyze this additional contribution by the Austrian economist more closely.

Generating Value in Innovation Processes

Schumpeter does not just distinguish invention from innovation. He also demonstrates that there is no innovation without diffusion, which normally takes place via markets. This idea is still widely accepted today. One must differentiate at least three phases in any innovation process: invention, innovation itself, and the later diffusion of this innovation in a city, economic sector, region, or country. According to their diffusion, then, one can distinguish innovations as local, regional, national, international, or nowadays global. Innovation studies must adapt their methodology to this scope of diffusion. Analyzing an innovation process at the local level is not the same as examining another that affects different countries and cultures, in which case one must employ methods with which to compare and verify the available data. Moreover, one must be able to collate this data. The *Oslo Manual* was created to satisfy those methodological requirements, as we will see later.

From Schumpeter's perspective, one must analyze this diffusion phase if one is to effectively understand the notion of creative destruction. If a business promotes and develops a specific innovation, such as synthetic silk for example, its competitors have no share in that new market, at least initially. So long as this innovation gains social acceptance, the synthetic silk market grows while that of natural silk stagnates or declines. Investors prefer to move their money into the new market, precisely because there are high expectations of growth there. In sum, in a few months (or years) competing businesses begin to show worse results and may even disappear if they do not react accordingly. The destruction of competing businesses, including their corresponding jobs, is one of the necessary consequences of a successful innovation. The extent of an innovation's success is measured during the diffusion phase. For that reason, Schumpeterian creative destruction finds authentic expression in this third phase of the innovation process. In the event of there being a prior equilibrium in the market, the innovation in question would break this, leading to an important change in relative values and hence losses and gains: "The effects of the entrepreneurial activity spread over the whole system, dislocating values, disrupting the equilibrium that existed before; the term Wind-fall correctly expresses the character of both these gains and losses."[24] For Schumpeter, this "dislocation" of values takes places in markets and creates gains for certain businesses (the innovative ones) and losses for others. For my part, I believe that innovation, in a philosophical sense, emerges through creating different types of values (social, political, cultural, esthetic, epistemic, and so forth) rather than through generating economic value alone. I will return to this point in chapter 5.

One of the qualities of the entrepreneur consists of knowing how to detect whether a new product will be received favorably or not by clients and consumers, as well as identifying what would be the best time to launch the product in the market. Commercialization is a fundamental part of this diffusion phase because if badly conceived it can lead to the failure of a product. There are numerous examples of this, of which I will mention one well-known instance: the struggle between Apple and Microsoft in the 1990s. Microsoft's more effective commercial strategy was successful, and Apple's operating system, even though better than Windows in the opinion of many experts, achieved less diffusion as a result. That decade, Microsoft won because it diffused its computer products better than Apple. Incidentally, Apple learned from that experience and nowadays pays extremely close attention to launching and commercializing its new products. Indeed, during the 2000s Apple was, thanks to the leadership of Steve Jobs, much more in-

24. Schumpeter, *Business Cycles*, 135.

novative than Microsoft and even overtook it in market capitalization.

Schumpeter examines many empirical examples, not just that of the silk market. He arrives at the conclusion that innovation is discontinuous and proceeds through a series of "explosions." According to him, it is "a disturbance of existing structures and more like a series of explosions than a gentle, though incessant, transformation."[25] This is one of Schumpeter's characteristic ideas, which may be valid as regards disruptive innovations but is more problematic when it comes to cumulative innovations.

In any event, economic wealth is both generated and destroyed throughout an innovation process. There are winners and losers. It is currently quite typical to equate innovation with the creation of value, and many authors define innovation in tandem with this feature. For my part, I will accept that innovation processes generate value but only if they are successful; in other words, if they overcome the diffusion phase. In other cases, they lead to devaluation; that is to say, they entail negative costs and the destruction of wealth for diverse agents and sectors. Creating value is one of the properties of innovation, but it is not sufficient in itself to define the concept.

I will end this brief excursion through Schumpeter's work by summarizing some of the results obtained. In *first place*, we now have at our disposal a definition inspired by his ideas, whose provisional formulation would be the following:

> Definition 3: *Innovations are new actions (and ways of doing), initiated by an entrepreneur, that transform X and generate value out of X, whether at the micro-, meso-, or macroscale.*

Innovative activities may take place in any economic sector. That is why I do not specify what X might be although we have already seen that, according to Schumpeter, innovation presents itself in the form of new products, new processes, new forms of organization in a business, new markets, and new primary materials. I would also say that X can likely be an economic entity but that there are other forms of innovation, such as social innovation.

25. Ibid., 100.

However, Schumpeter does not take this into account. As a result, this third definition is still insufficient for my purposes, and I will have to modify it at a later stage. Its main defect is that it does not take into account the evolving nature of innovation that entails the intervention of several agents, whether economic or of another variety.

In *second place*, one must differentiate between originality and innovation. Not all new actions are innovations; the important thing is that they are valued positively, for example in markets. I will return to this point in chapter 5.

In *third place*, entrepreneurs innovate and innovations are produced in markets. I should say here that I consider this the most contentious point in the Schumpeterian notion, as I will attempt to demonstrate in chapters 2 and 3.

In *fourth place*, Schumpeter is conscious of the fact that not all innovations are successful. It is important to remember this because many ideologues of innovation argue that something is only innovative when it is successful in markets. This is not the case. Innovative ideas have often failed down through history. And it's not just ideas, given that innovation processes do not just involve ideas. As Henry Chesbrough observes, "Most innovations fail. And companies that don't innovate die."[26] One must, therefore, pay attention to failed innovations; not just their successful counterparts. Learning from one's own mistakes is a key quality of innovative people.

In *fifth place*, an innovative agent creates but at the same time also destroys value. An innovative entrepreneur can do a lot of harm to other firms and even force them to close down for not being competitive enough, with the concomitant effects for both workers and shareholders. Economic innovation processes in free markets imply creation in some areas and destruction in others.

In *sixth place*, there are not just different types but also different scales of innovation. This is very important for innovation studies given that the methodology will change according to the case study: in other words, whether local, urban, regional, national, or global, and whether at the micro-, meso-, or macroscale.

26. Chesbrough, *Open Innovation*, xvii.

The Emergence of R&D Systems

Although Schumpeter contributed the first theory of innovation, his initial influence was mostly in academic circles. His ideas only began to have an impact on innovation policies in the 1980s, when authors such as Richard R. Nelson, Sidney G. Winter, Giovanni Dosi, Christopher Freeman, Peter F. Drucker, and others reclaimed the ideas of the Austrian economist and proposed them as a suitable conceptual framework for interpreting the profound changes resulting from the new ICTs and the diverse disruptive innovations generated by technoscience during the final quarter of the twentieth century. During the previous age (1945–1980), the dominant paradigm in matters of innovations was the linear model, whose main ideas were sketched out by Vannevar Bush in his celebrated report to President Roosevelt, *Science, the Endless Frontier* (1945). According to this report, the main disruptive innovations that had been generated during World War II (nuclear energy, radars, computers, penicillin and other pharmaceutical products, the artificial silk industry, and so on) originated in scientific research and in certain technological developments that emerged as a result of such research. According to Bush, the entrepreneur who seeks to innovate should collaborate with scientists and engineers, and even create a R&D department in his own company. Although firms and industries are the innovative agents par excellence, teamwork among scientists, technologists, and entrepreneurs is the fastest way to generate innovations. The consolidation of national patent systems, especially in the United States, was a key factor in the growth of R&D. Since then, this has been central to public innovation policies as well as many business strategies.

The Report *Science, the Endless Frontier* by Vannevar Bush

Vannevar Bush, through his report *Science, the Endless Frontier* (1945), is usually attributed with designing the first US science, technology, and innovation system. Despite this, some authors such as Daniel S. Greenberg have ridiculed this myth of a "creator" of the new science and technology policy.[27] Prior to publication of the report, some American universities like MIT (the Massachu-

27. Greenberg, *Science, Money, and Politics*, chap. 3.

setts Institute of Technology), Berkeley, and Stanford had taken steps in that same direction. However, the establishment of a new model to manage science throughout the country as well as Congress approval for this in the postwar era could be considered a political innovation that had important ramifications for the development of science, technology, and innovation all over the world through the twentieth century. Bush worked for the White House during World War II as director of the Office of Scientific Research and Development, and from this position he made a number of original proposals. Furthermore, he managed to get them carried out in practice, although not without some difficulty. The consequences derived from creating institutions with a new imprint, such as the National Science Foundation (NSF), were so important that they transformed Bush into one of the great political entrepreneurs of the twentieth century (here I use the term "entrepreneur" in the same way as Schumpeter). In arguing this, I am applying one of the central trains of thought in the present work: namely, my aim to reinterpret certain Schumpeterian ideas and project them onto other spheres of human action – in this case political activity – instead of restricting them to business innovation alone. Neither firms nor the private sector have a monopoly on innovation; it also emerges out of the public sector, as we will see in chapter 4.

Of course, Bush was completely aware of the originality of his report, as he pointed out to President Roosevelt:

We have no national policy for science. The Government has only begun to utilize science in the nation's welfare. There is no body within the Government charged with formulating or executing a national science policy. There are no standing committees of the Congress devoted to this important subject. Science has been in the wings. It should be brought to the center of the stage – for in it lies much of our hope for the future.[28]

These arguments are extremely interesting because they call for the creation of a standing committee on science in Congress and a federal agency, which would in time become the NSF. Moreover, they propose creating new institutional agents: federal agen-

28. Bush, *Science, the Endless Frontier*, 13.

cies specializing in R&D. With time these innovative proposals were successful but they initially encountered opposition, as is often the case with almost all innovations. Defenders of the previous tradition (that of the priority of individual states over the federal government, a mistrust of government intervention in scientific issues, searching for private patrons to help universities and research centers, and so on) put up considerable resistance to these new ideas. The result of this was that Bush needed strong political support in order to get his proposals across. Even so, it took President Truman four years to implement them. One should also underscore the fact that, although Bush was a very prestigious scientist, his proposal did not originate in any laboratory but in a different space altogether: the White House Office of Science Policy. Likewise, one should bear in mind that although the report was drawn up and sent by Bush, numerous other people – scientists, engineers, entrepreneurs, heads of research centers, university presidents, and qualified representatives of scientific communities – also collaborated in the task. This was, then, a politico-institutional innovation promoted by a leader but designed and agreed on in tandem with an important group of experts coordinated by Bush. Incidentally, Bush was not just a first-rate researcher but he also had prior experience in managing scientific and academic institutions.[29]

Ultimately, the Bush report was fully embraced politically after multiple controversies, debates, and critiques that continued well after the war had ended. Not everything proposed by Bush became reality and many other things were implemented besides what he had suggested. Nevertheless, his report was the trigger that led to a profound transformation in American science and technology, a transformation that could be summed up in the following

29. Bush came from MIT, where he was not just a prestigious scientist but had also become vice president of the university and dean of the School of Engineering in 1932. In 1938, he moved to Washington, DC, to become president of the Carnegie Institution and during World War II became director of the Office of Scientific Research and Development (OSRD) in the White House. Note that the term R&D (scientific research and development) was already incorporated into the acronym for this latter institution.

terms: *in the 1950s the United States created a R&D system.* As this new system grew and developed so did future leading institutions in the emerging technoscientific field, such as the National Science Foundation (NSF), the National Institutes of Health (NIH), and the National Aeronautics and Space Administration (NASA).[30] If we liken the creation of these institutions to that of the Royal Society in seventeenth-century England and compare their respective objectives and statutes, we get an impression of the radical change experienced in the United States in the aftermath of World War II: namely, the first phase of a technoscientific revolution that was characterized by the emergence of Big Science.

It is worth citing certain passages from the Bush report. In a letter dated November 17, 1944, President Roosevelt asked Bush to answer four specific questions: (1) What can be done, consistent with military security, and with the prior approval of the military authorities, to make known to the world as soon as possible the contributions which have been made during our war effort to scientific knowledge? (2) With particular reference to the war of science against disease, what can be done now to organize a program for continuing in the future the work which has been done in medicine and related sciences? (3) What can the Government do now and in the future to aid research activities by public and private organizations? (4) Can an effective program be proposed for discovering and developing scientific talent in American youth so that the continuing future of scientific research in this country may be assured on a level comparable to what has been done during the war?[31]

Using this presidential guide as his starting point, Bush answers with focus on the natural sciences, including biology and medicine. His main argument is expressed in the letter accompanying

30. In 1947, by which time Bush had been replaced by John R. Steelman, the Scientific Research Board was established in order to advise the US President on science policy. Because Bush did not include medicine in his report, the National Institutes of Health, which were at the time not so important, chose to create the National Foundation for Medical Research, which would make a decisive contribution to transforming medicine into technomedicine. There were similar developments in other fields of knowledge.

31. Bush, *Science, the Endless Frontier*, 3.

the report: "Scientific progress is one essential key to our security as a nation, to our better health, to more jobs, to a higher standard of living, and to our cultural progress."[32] When Bush speaks about scientific progress here he is referring to advancing knowledge, in other words achieving *new knowledge*. This basic or fundamental idea is expressed clearly in the initial sentences of the report summary, or that point in the document most readily consulted by the likes of President Roosevelt and members of Congress:

Progress in the war against disease depends upon a flow of new scientific knowledge. New products, new industries, and more jobs require continuous additions to knowledge of the laws of nature, and the application of that knowledge to practical purposes. Similarly, our defense against aggression demands new knowledge so that we can develop new and improved weapons. This essential, new knowledge can be obtained only through basic scientific research.[33]

Bush was, then, one of the first people to contend that knowledge was not just a source of power, an idea already conceived by Francis Bacon in the seventeenth century, but also of economic wealth, social progress, and better health. In the previously cited passage he alludes constantly to new products, new industries, new arms, new medicines, and so on. And the important point here is that, according to Bush, all these potential innovations are only possible if, beforehand, scientists have developed *new knowledge*. Although he does not state it explicitly, he suggests that any form of innovation (or at least the most significant innovation) is only possible thanks to generating new knowledge and more specifically new scientific knowledge. Herein lays the conceptual nucleus of the linear theory, which can be summarized as follows:

1. It will always be possible to generate new scientific knowledge. Science is an endless frontier.
2. The principal innovations (technological, military, business, industrial, health, and so forth) depend on such advances in scientific knowledge, to which (foreshadowing my own proposals) I would add what I will term *techno-*

32. Ibid., 4.
33. Ibid., 6.

scientific innovations as well as non-scientific discoveries. Science and technoscience differ from one another for a number of reasons, among which is the fact that the former produces discoveries (new facts, new hypotheses, new laws, new theories, and so on) while the latter generates innovations. Bush always speaks in the name of science, yet his report heralded the Age of Technoscience.

3. Scientific knowledge is knowledge par excellence, especially knowledge of the natural sciences, which are those that inquire into the rules of the natural world.

These three postulates imply further assumptions that I will address in the following pages. For the moment, I want to underscore the fact that these initial sentences written by Bush in his report constitute a significant contribution to the theory of innovation: they make innovation processes depend on knowledge and, more specifically, scientific knowledge. Nowadays, almost all the authors who discuss knowledge society accept this idea: namely that innovation depends on knowledge. Manuel Castells, for example, closely links the two concepts: "knowledge and innovation are the sources of wealth, of power, and of quality of life."[34] One could cite other authors who defend the same thesis, including those of numerous official EU documents.

Schumpeter, then, proposed innovation as a source of economic wealth, and some years later Bush suggested a new idea whereby scientific knowledge underpinned significant innovations. To my mind, this close relationship between scientific knowledge and innovation is one of the pillars of the technoscientific revolution,[35] although with the following qualification: in the case of science, the search for knowledge is an end in itself; in the case of technoscience, advances in knowledge are important, but they are just a means to an end rather than an end in itself. The contemporary technosciences (techno-mathematics, techno-physics, techno-chemistry, techno-biology or biotechnologies, techno-geology, techno-astronomy, techno-pharmacy, techno-medicine, and the techno-social sciences) closely link innovation and knowledge,

34. Castells et al., *La transición a la sociedad red*, 20.

35. See Echeverría, 2003.

but with the latter subordinate to the former. Since their emergence, R&D systems have tended to invoke innovation as one of their main objectives. That is why they later came to be officially termed RDI systems.

Returning to Bush's report, right from the outset he also clearly states the second main point of his proposal, which was also original at that time: "Science can be effective in the national welfare only as a member of a team, whether the conditions be peace or war."[36] He thus addresses what had been typical US practice during the war at a theoretical level: in developing confidential research projects, scientists had collaborated closely with engineers, the military, industrialists, and politicians, leaving the traditional ivory tower of academic science. That teamwork implied an important change in scientific practice in that it was no longer a question of forming teams from the same academic discipline. Cooperation was not just interdisciplinary (scientists and engineers, for example), but "intersectorial" with team members coming from very diverse professional sectors, many of which were not academic. This transformation of scientific practice prefigured the linear model, in which three consecutive phases were assumed in any technoscientific project: first, scientists would generate knowledge; then, engineers and technicians would convert that knowledge into technological prototypes; finally, entrepreneurs would select some of these prototypes and produce them en masse, after previously checking their effectiveness, efficiency, and if applicable, utility for potential clients and users, some of whom were federal government institutions and agencies. One of the priorities in this case was military innovation, something that obviously gains in value during times of war in order to reduce one's own casualties and increase those of the enemy. Put in contemporary terms, the linear model and the transfer of knowledge and information among scientists, engineers, and entrepreneurs (in that order) defines the basic strategy that one must follow to encourage business innovation. Most disruptive innovations dating from that era emerged thanks to scientific and technological research. First of all, the technoscientific revolution is characterized because scientific and technological

36. Bush, *Science, the Endless Frontier*, 6.

knowledge is the basis of wealth, power, and well-being and also because it generates disruptive innovations. For that reason, it is necessary for scientists, technologists, and entrepreneurs to work in teams, under the watchful eye of those who commission and finance the major research projects: above all, government agencies.

These are Bush's two main responses to Roosevelt's four questions: first, new knowledge is indispensable when it comes to innovating; second, scientists, engineers, and entrepreneurs must work together as a team. Once the war ended, the federal government should resolutely promote scientific and technological research and contribute financial resources, as well as research institutions and infrastructures, to this end. Scientists, technologists, entrepreneurs, politicians, and military personnel had collaborated actively during war and produced important innovations, some of which were decisive in the United States winning the war. However, this had only happened in areas of interest to the Department of Defense. During the postwar era that policy would have to be extended through government financing of the complete spectrum of the natural sciences and planning of large-scale research projects. Bush suggested to Roosevelt that the government should become the main technoscientific agent in the country. And, although with some reticence, that is indeed what it did. Bush was therefore successful in his capacity as a politico-institutional innovator. The US government and Congress created science policy commissions and set aside funds to promote research and development. The NSF was also established, followed by other federal technoscientific agencies, both military and civil. The nascent American R&D system began to grow and strengthen. Shortly thereafter, other countries began to imitate this system, all of which guaranteed the diffusion of that original disruptive innovation. Not for nothing it produced a technoscientific revolution that shaped our era, which we might justifiably term the Age of Technoscience.

In addressing defense spending, Bush demonstrates that modern warfare is "a battle of scientific techniques,"[37] citing the examples of the battle against German U-boats and the development of the radar and other new weapons. The defense and attack capabil-

37. Ibid.

ities of a nation, he contends, depend strictly on scientific knowledge. For that reason, he concludes, "There must be more – and more adequate – military research in peacetime."[38] He explains this idea in more detail: "This can best be done through a civilian-controlled organization with close liaison with the Army and Navy, but with funds direct from Congress, and the clear power to initiate military research which will supplement and strengthen that carried on directly under the control of the Army and Navy."[39] The armed forces will maintain their own research centers, but he also proposes creating an organization to coordinate links between civilian and military bodies, which would be financed by Congress. Bush emphasizes once more that it should be the US government and Congress that lead the transformation he is proposing, but also that the autonomy of the military agencies that sponsor scientific and technological research should be maintained.

Industrial spending completes the design of this new science and technology system:

> To create more jobs we must make new and better and cheaper products. We want plenty of new, vigorous enterprises. But new products and processes are not born full-grown. They are founded on new principles and new conceptions which in turn result from basic scientific research. Basic scientific research is scientific capital. Moreover, we cannot any longer depend upon Europe as a major source of this scientific capital. Clearly, more and better scientific research is one essential to the achievement of our goal of full employment.[40]

The syllogism here is blunt and at the root of the linear model. Observe how Bush uses very precise terminology in the sense that he is speaking about new scientific knowledge as a means of generating new products and new processes, in other words, innovations along the lines of the terminology used by Schumpeter, who had published some of his most influential works between 1939 and 1942. Full employment and the progress of a society cannot be achieved without competitive firms. These firms are not competitive if they are incapable of making and selling new and cheaper

38. Ibid.
39. Ibid.
40. Ibid.

products. However, commercial and industrial innovations only emerge if there are technological developments and scientific advances. For that reason, scientific research is the foundation of innovation, of business progress, and full employment.

This first outline of the American R&D system produced profound changes in scientific practice, which soon became closely linked to industrial activity on the one hand and military agencies on the other, as well as to the norms and priorities established by government agencies and the White House itself. Among all these transformations, I will highlight three in particular:

1. Considerable budgetary allocations to encourage R&D. Federal aid grew 14 percent annually in constant dollars between 1953 and 1961. As Bruce L. R. Smith points out, "Growth was both a condition and a part of the doctrine,"[41] because it allowed for the silencing of discrepant scientists and the favoring of their more compliant counterparts.
2. The creation of commissions to draw up science policies and decision-making in Congress. Thus a new kind of technoscientific agent emerged: advisory committees, which are always made up of experts in the material in question and which have gained increasing importance in R&D systems because these committees recommend certain decisions. They are not quite as powerful as a company's board of directors, but they are very influential when it comes to defining innovation policies and strategies, both in public agencies and private firms. Scientists (as well as technologists and entrepreneurs) frequently fight one another to get onto these committees and expert groups. One of the consequences of the new R&D system was the emergence of lobbies and interest groups, some of which were very influential in designing and applying public RDI policies.
3. In the American case, the NSF and other federal agencies taking on the task of promoting, financing, and developing diverse strategic projects. For that reason, many of the innovations stemming from the nascent R&D system emerged out of technoscientific agencies' strategic deci-

41. Smith, *American Science Policy since World War II*, 39.

sion. This is how the tradition of priority research areas, which have influenced contemporary technoscientific activity in a major way, began. In the same way as the US federal government and Congress, different federal agencies defined their objectives, indicated and financed their main areas of research, and developed specific programs to reach those objectives. On some occasions, these very agencies themselves designed major research programs: the Manhattan Project, the ENIAC project, the project to land on the moon, the Hubble Space Telescope, the fight against cancer, the Mars Exploration Program, the Human Genome Project, and so on. In the Age of Technoscience promoted by governments, the most important advances in knowledge and principal innovations were produced in those areas that had been previously selected as priorities. This was guided, directed, and even planned progress.

4. Boosting research projects and contracts and making research teams compete at the federal level. This forced scientists to outline what they wanted to do beforehand (the state of the issue to be researched, a starting hypothesis, objectives to be achieved, the timeframe for this, the necessary human and economic resources, the methodology to be applied, a work schedule, scientific advances, technological developments, any possible expected innovations, and so on). If they wanted to receive additional funding, universities and research centers had to adapt to government policies as well as to the research priorities of the NSF and other agencies. Not without some disapproval, research became strongly influenced by R&D policies.
5. The emergence of new professions such as R&D advisors and management experts, as well as specialists in evaluating science and technology. These agents have gained increasing influence as the technoscientific system has become more consolidated, leading to considerable bureaucracy in R&D management.
6. The introduction of indicator systems to measure R&D activities. In this field, the main reference point is the

Frascati Manual, which I will discuss later, together with the *Canberra Manual*, which focuses on human resources linked to R&D. A second reference point is made up of impact factor indicators in scholarly journals, which has become the principal method for measuring excellence in research.[42] I will examine these in more detail below when I discuss scientometrics.

7. The fomenting of patents and knowledge transfer to industry through economic incentives. Roosevelt's question to Bush about how to spread the contributions of secret knowledge during the war publically hints at a key theme in technoscience: the diffusion and divulgation of scientific knowledge, of technologies, and of innovations.

Two Major Technoscientific Innovations: The Atomic Bomb and the ENIAC Computer

Here I will discuss two major examples of Big Science that were developed during World War II in the United States and that led to disruptive innovations that marked the beginning of the Age of Technoscience. They are both good examples of Bush's linear model, and advances in scientific knowledge and the research of engineers and technicians were also both indispensable for bringing them about. Moreover, both projects were sponsored and financed by the US Department of Defense, given their strategic importance. In discussing these two examples, I seek to highlight the existence of an important technoscientific field, military RDI, which is often overlooked in innovation studies despite its huge economic, technological, and military importance. The Schumpterian innovative entrepreneur may be a physical person, but could also be embodied by a military institution and a group of people that develops a secret RDI project in times of war, as was the case with the Los Alamos National Laboratory and the ENI-

42. The most influential of these was the ISI system, which was later termed ISI-Thomson and is now known as Thomson-Reuters. Several alternative systems have emerged in Europe in recent years with varying degrees of success. At the time, the ISI system was an important technoscientific innovation that, understandably, subsequently gained competitors.

AC project at the University of Pennsylvania's Moore School of Electrical Engineering. Any theory that aspires to be a valid means of analyzing twentieth-century and early twenty-first-century innovation policies and innovative practices cannot afford to ignore the importance of military areas for technological innovation; the main goal of such areas, of course, as Bush indicates in his report to President Roosevelt (*Science, the Endless Frontier*) is to design and create new weapons. Just as science contributes an endless frontier and facilitates the continual emergence of innovations, so producing new weapons for both defense and attack purposes is likewise limitless. The example of the atomic bomb is particularly illustrative because it tragically exemplifies the Schumpeterian notion of creative destruction. It is a similar case with the ENIAC series computers, which in principle were simple calculation instruments for military purposes but, with time, were converted into assessing all fields of human action rather than just being limited to military activity. Both (especially the Manhattan Project) were canonical ventures for Big Science. They also both demonstrate that major innovations contribute new capacities for action; not just new knowledge and new products. For all these reasons, it is worth paying attention to these two case studies that both effectively illustrate Bush's thesis on scientific knowledge as the driving force of innovation.

Big Science first emerged in the world of physics, with its different radiation laboratories, the Manhattan Project, and the construction of cyclotrons. Physics was also central to the restructuring of scientific activity promoted by Bush and his successors toward US scientific policy. The macro-projects of the 1940s and 1950s had a strong industrial dimension and were developed through the collaboration of scientists, engineers, and entrepreneurs in accordance with the "science push" innovation model, which could also be termed STI (science, technology, and innovation). As regards links with science policy, many prestigious physicists were involved in this, beginning with Vannevar Bush himself. A new profession thus emerged: scientific and technological knowledge management, which is nowadays an established career in both the public and private sectors.

This change in the structure of scientific practice did not just affect people but also institutions. Different (and in general the most important) departments and research centers of applied physics collaborated closely with American companies and industries. Examples of Big Science abound in mid-twentieth-century physics, both in the United States and other countries, although here I will just concentrate on the Manhattan Project because it was the most important of all for many reasons.[43] There is still no in-depth study of the physics technosciences during the twentieth century, although there have been important contributions to this end.

In any event, the Manhattan Project is the best example of Big Science in the field of physics, which is why I will limit myself to discussing this. From a financial perspective, it was the first macro-project of the twentieth century and only surpassed by the great space exploration projects to come. It has been calculated that, during the period 1942–1945, the United States invested two million dollars into this project (in other words, the equivalent of around one hundred billion dollars at the end of the twentieth century). It was a top-secret military project that was fully supported by the White House. Theoretical and experimental physicists collaborated with chemists, mathematicians, engineers, and numerous industries on the project. Developed at a dizzying rate, it constitutes an excellent example of disruptive innovation based on science and technology. It is worth underscoring the fact that both its design and its objectives experienced significant changes

43. In the case of Europe there were institutions that opted for a close collaboration between scientists, technologists, and entrepreneurs. A pioneer in this was the Dutch organization TNO (The Netherlands Organization for Applied Research), which had already been founded in 1932. Established in Bavaria in 1949, the German Fraunhofer Society, which specialized in applied research, was also important. The Max Planck Institute, however, continued to focus on fundamental science. None of these institutions promoted Big Science, in contrast to their counterparts in the United States. The first major Big Science project in Europe was the CERN (Conseil Européen pour la Recherche Nucléaire, European Organization for Nuclear Research), founded in 1954 by twelve European countries. Here I will not discuss the development of Big Science in Europe or the Soviet Union.

as a result of the scientific discoveries that were made during the actual research process. From the outset, work was carried out in two directions: one geared toward the fabrication of a uranium or plutonium fission bomb and the other toward an implosion bomb. At first, the fission bomb took precedence. In 1944, priority was given to the plutonium bomb and research began on thermonuclear bombs, the possibility of which had been suggested by Edward Teller, one of the physicists involved in the project. The end results were most satisfactory for those who promoted the project and for many of those involved in its development. Scientific knowledge advanced enormously, and it led to numerous technological developments and innovations. The most important of these was the discovery of a new source of energy, nuclear energy, the military, political, technological, and commercial control of which then became a new objective over which different states and major companies began to compete (without forgetting the competition among scientists, engineers, and technicians). Some firms made millions of dollars thanks to the Manhattan Project. The US military achieved an undeniable success, which left its mark on army strategies throughout the twentieth century: indeed, possession of the atomic bomb became the goal of all the great powers. In short, from the perspective of those who promoted the project from the outset and the different groups that participated actively in it, the results were satisfactory. The Manhattan Project was the first major example of a technoscientific macro-program and established a standard for the subsequent development of technoscience, both in physics and in other disciplines. As Bush had hoped, the advance in knowledge gave rise to new technological devices and, in the final analysis, major innovations, such as atomic bombs and nuclear power plants. In line with what Schumpeter had contended, this innovation both altered the global energy production system and radically transformed the idea of war (through the emergence of the "Cold War" concept) as well as military defense systems (through, for example, the creation of antinuclear missile shields).

However, all of this also drastically increased humanity's destructive capacity, with all the dramatic consequences this implied. The human, social, and environmental damage that the Hiroshima and Nagasaki bombs did in Japan went beyond anything done by any other weapon of war in history. And the subsequent development of nuclear weapons by the Soviet Union led to the "balance of terror," an unprecedented situation in history given the potential scale of disaster had a nuclear war ever broken out. Moreover, the new problem of nuclear waste emerged, which continues to this day without being fully resolved. And one should not overlook the issue of just how safe nuclear power plants are, which was raised tragically once more recently following the earthquake in Fukushima, Japan, in March 2011. The Manhattan Project is a canonical example of creative destruction in the Schumpeterian sense of the term, and not just because of the Hiroshima and Nagasaki bombs, but also as a result of the new problems deriving from the use of nuclear energy. There is no doubt that it brought with it enormous opportunities, gains, and benefits but also risks, losses, and damage. In particular, it significantly modified humanity's scale of values, to put it in Nietzsche's terms. Moral and religious valuations of the bombing of Hiroshima and Nagasaki were mostly negative, with a few rare exceptions. However, these did not figure into the course of military action taken at the time, nor even thereafter, when it came to the issue of whether to slow down the arms race. In fact, this objective was only entertained many years later, once various states and many companies had come to dominate the nuclear technosciences. The Manhattan Project is a clear example of the de facto primacy of military, political, economic, and technological values over moral, religious, environmental, and in many cases, social and cultural values. Few examples demonstrate so clearly the fact that macro-science and technoscience have produced new value systems, such as the balance of terror. In sum, the Manhattan Project is a canonical illustration of the new technoscientific paradigm. Human beings had never previously demonstrated so great a capacity to transform the world around them, in this case destructively. Thanks to a set of disruptive innovations that derived from the Manhattan Project, contemporary science and technology have demonstrated their capacity to both create and destroy. The project is also a prime example of Schumpeter's creative destruction, although I am unaware if Schumpeter ever concerned himself with this shining illustration of his theories. One could say the same for Bush, who also guessed correctly in laying the foundations of the linear model of innovation. The usefulness of basic or fundamental scientific research, conveniently implemented by the rest of the team (engineers, technicians, entrepreneurs, military personnel, and financiers), was demonstrated fully in the development of this macro-project, and especially in its results. However, since it was also demonstrated that innovations lead to minuses as well as pluses, any valuation of innovation processes must be consequentialist, measuring not just the results but also any later consequences.

The Los Alamos National Laboratory was a military institution in which certain scientists were very influential. It was the first practical example of the potential effects of the so-called "social contract" among politicians, scientists, engineers, industrialists, and the military. Internally, the structure of its different departments was very hierarchical. J. Robert Oppenheimer, the laboratory's director, had the authority to transfer scientists from one team to another, according to project priorities. Scientists and engineers collaborated closely within each team. There was a governing council and a coordinating council to overlook the work of the different departments, of which there were originally four: the Theoretical Division, the Chemistry and Metallurgy Division, the Experimental Physics Division, and the Engineering and Ordnance Division.[44] These titles left no room for doubt as to the military nature of the research, as did the scientific and technological personnel involved. General Leslie Groves, director of the Manhattan Project, and Oppenheimer pursued a flexible strategy right from the outset: although priority had been given to researching fission, other areas of research were also encouraged so long as they might lead to the creation of an atomic bomb that, should this happen, could be mass-produced. As regards this second objective, plutonium seemed preferable to uranium because it could be made in a nuclear reactor, which the private firm Du Pont was hastily charged with constructing in Chicago. This major feat of

44. These were the divisions located in Los Alamos. Other divisions were located elsewhere in the country, such as the Medical Division.

engineering was crucial for the development of the project because the phenomenon of spontaneous and relatively frequent (once a month) fission of uranium was discovered at this nuclear reactor, with this becoming a new subject of inquiry for scientific research. The industrial production of plutonium, which was necessary for the Manhattan Project, in turn facilitated the study of a new and until that time unknown physical concept.[45] At the very least, the research priorities of the Manhattan Project changed. The small group that had been researching the implosion bomb grew from a total of five to fifty members in a few months.

As experts were needed in explosives (George Kistiakowsky), fluid dynamics (John von Neumann), computing (IBM, ENIAC), and spontaneous fission (the Berkeley group, directed by Emilio Segrè), they were immediately signed up for the program.[46] In the meantime, chemists worked on separating plutonium-240. Another feature of the Manhattan Project was the similarity that stretched across different areas of research. While important issues still remained to be resolved (whether theoretical, instrumental, those concerning the industrial production of plutonium, those concerning prior testing, and so on), the remaining teams continued independently with their research. Effective coordination of the different stages and foreseeable achievements was essential for the success of the project, as well as flexibility when it came to changing priorities and maintaining alternative areas of research in tact just in case something went wrong. The final and strictly military objective took precedence over how far the scientists involved in the project believed in the most appropriate theoretical assumptions and the seemingly most promising areas of research. For that reason, the Manhattan Project is also interesting from an organizational perspective and not just as a product innovation (the atomic bomb). I could keep on analyzing many other aspects of the project but my discussion thus far will suffice for this section, which is limited to offering the first major example of Big Science.

45. I use Lilian Hoddeson, "Mission Change in the Large Laboratory," 275 as the basic reference point for my discussion here.

46. One should not forget, however, that some scientists refused to participate in the Manhattan Project and were even involved in the establishing the Pungwash movement that opposed atomic bombs.

One should also bear in mind, though, that as well as designing the bomb and demonstrating its theoretical possibility, it would also have to be built and tested.[47] One thing is producing a first prototype (R&D) and another is industrial innovation properly understood, which implies the mass production of a new product or in this case the emergence of a whole new industrial sector, the nuclear industry. In Schumpeterian terms, here we are witnessing the appearance of a new primary material (a new form of energy) as well as the creation of a new market (that of nuclear weapons and nuclear energy), both of which have grown since that time. In sum, then, the Manhattan Project was an innovation in the five senses in which Schumpeter discusses innovation, including the already mentioned notions of innovation as a new kind of research (process innovation), teamwork, and simultaneity. The Manhattan Project represents one of the great paradigms of macro-science, but also of innovation. Moreover, it is a canonical case based on the linear model given that, without any of these advances in the search for knowledge – which did in effect take place – nothing would have been possible. The making of the atomic bomb was an issue of major technological, industrial, and financial complexity, but one must remember that the client that had commissioned it (the US Department of Defense) ended up using the innovation, both at an experimental level and later (and effectively) in the field of battle. One of the most notable features of this case was the plurality of agents needed to carry it out, together with the primacy of military objectives over any others, including advancing knowledge and technological developments. Military R&D is characterized, logically, by being at the service of military strategies and objectives. For that reason, neither innovation in itself nor the quest for knowledge is the ultimate goal. In any event, nothing would have been possible were it not for that teamwork about which Bush spoke in his report to President Roosevelt in 1945, published just

47. The date set aside for the first experimental test, May 1, 1945, was key in planning the activity of the different teams during the final stage of developing the project. With this, it was clear that the production of knowledge had been planned out in industrial terms, in contrast to traditional scientific research. In fact, the first test took place on July 16, 1945.

as the Manhattan Project was reaching its culmination. As a result, the Schumpeterian model based on the innovative entrepreneur is called into question by the example of the Manhattan Project. Innovations based on R&D involve multiple agents, including of course entrepreneurs but also scientists, engineers, politicians, and last but not least, innovative military personnel. This is an example of plural innovation, innovation undertaken by several innovative agents, each of whom has their own cultural background. I will return to this issue when I advocate a pluralist innovation studies model.

From a methodological point of view as well, there was considerable pluralism in this case. The success of the Manhattan Project cannot just be attributed to theoretical changes (although there were some) or new scientific methods (of which, too, there were some). Rather, what changed dramatically throughout the course of the project was scientific and technological practice itself, and this due to a number of very different reasons. Without huge funding on the part of the US government, a military dimension being applied to research, the collaboration of major industrial companies (some of which had incorporated new methods of production and organization), and the strict discipline to which scientists and engineers were submitted, it would have been impossible to make the atomic bomb in such a short space of time. The urgency stemming from the wartime context was decisive when it came to the success of this disruptive military innovation. Later, the Manhattan Project came to be considered an exemplary illustration of technoscience, as well as RDI in the military field. For that reason I consider it one of the foundational moments of the Age of Technoscience.

An examination of other examples of macrophysics such as the Brookhaven National Laboratory or the CERN would lead us to the same conclusions. Such a detailed study of the physical technosciences would, though, require much more space as well as additional analytical perspectives. Instead, let me offer a final observation on this point: technologists and scientists from very diverse scientific disciplines, including the then-emerging computer sciences, took part in the Manhattan Project. Calculating the mass of enriched uranium or plutonium, which was necessary to explode the atomic bomb, would have been impossible without the aid of IBM computers and the new techniques of numerical analysis that these machines offered, in the same way as ENIAC during the project's final phase as regards simulations in fluid dynamics. Technoscience does not just generate discoveries and new theories; it also produces technological developments and innovations. The key aspect is that it should generate them from a foundation of previously established research practice that requires profound hybridization among scientists and technologists, as well as with the help of other technoscientific agents with their own objectives and goals.

Let us now consider a second example: the ENIAC (Electronic Numerical Integrator and Computer) project, which inaugurated what I will term techno-mathematics in order to distinguish this from the traditional discipline of mathematics. An important disruptive innovation was also produced as a result of this project: computers. In general terms, numerous computers were constructed and for different aims. Let me briefly survey how all this came about.

In 1930, Vannevar Bush built a differential analyzer designed to resolve important equations in the study of electrical circuits. Then the German Konrad Zuse devised a universal calculator, the Z3, which he completed in 1941. It was a small machine with a punch tape reader, a console for the operator, and two units with 2,600 relays that could do several mathematical operations such as multiplications and square roots of numbers with 22 bits. One had to introduce the initial values by hand and as such it displayed important technical deficiencies. Scientists, but also military personnel, used these machines: for example, Zuse's Z4 was used in 1943 in operations against Allied ships in the Mediterranean.[48] Another major project was the Mark I, initiated by Howard H. Aiken at Harvard in 1937. Its principal innovation consisted of introducing a data register that would later become computer memory. Financed by IBM, the Mark I was presented in 1944 and

48. In this section, I take my information from Breton, *Historia y crítica de la informática*, 69–81.

offered immediately to the US military for its calculating power. These were all electromechanical machines. Another innovation, vacuum or electron tubes – used in the Atanasoff-Berry Computer (ABC) pioneered by John Vincent Atanasoff and Clifford Berry in the late 1930s and successfully tested in 1942 – facilitated the creation of the first electronic calculators as well the digital (as opposed to decimal) representation of numbers. In sum, from the 1930s on, a number of wide-ranging innovations were introduced that, during World War II, resulted in the design and fabrication of the first computers made in accordance with von Neumann's architecture. And these innovations increased humanity's computation capacity; in other words, they generated new actions (processing non-linear equations, the computation of numbers inaccessible to the human mind, and so on) and ways of doing (faster and more accurate calculations), radically transforming the field of mathematical actions. It is typically assumed that mathematics is a theoretical and abstract science, and indeed it is. However, mathematics is also characterized by a series of actions that are specific to mathematicians, such as arithmetic operations, solving linear equations, constructing geometric curves and figures, solving differential, integral, and statistical problems, and so on. The emergence of techno-mathematics, as I term computer sciences, implied a radical innovation in the history of mathematics as a whole, given that it entailed a new form of mathematics that cannot be practiced without the aid of computers. These early techno-mathematical artifacts, together with their predecessors, from the seventeenth-century arithmetic machines of Blaise Pascal and Gottfried Wilhelm Leibniz to Charles Babbage's nineteenth-century "analytical engine," helped implement classic mathematical actions in another way and also facilitated new mathematical actions. They were, therefore, mathematical innovations in the Schumpeterian sense and I will treat them as such here.

The first major example of an operative techno-mathematical artifact was the ENIAC, developed at the University of Pennsylvania's Moore School of Electrical Engineering from 1943 on. This was also a secret military project ("Project PX" at the US Army's Ballistic Research Laboratory). It had started out as a technological project, but during its later stages it was joined by the team headed by von Neumann, one of the vanguard researchers in mathematical logic and other branches of mathematics. Von Neumann contributed a lot of theoretical knowledge and directed the team toward basic or fundamental scientific research. The team was also made up of John Mauchly, who had been appointed by the military for the project, Herman Goldstine, and J. Presper Eckert, an engineer who managed to integrate several technical improvements that had been developed in the 1930s in the emerging field of computer sciences into the ENIAC prototype. The first ENIAC prototype consisted of 17,648 vacuum or electron tubes, 70,000 resistors, 10,000 capacitors, 1,500 relays, and 6,000 manual switches, making it a very large and highly complex machine whose operating required considerable technical abilities. It cost a fortune, $500,000 at that time, but functioned at great speed and was at the same time programmable and universal; in other words, it was applicable to different types of calculations. Its electricity consumption and heat emission were both enormous and as a result it had to be continually cooled down. Following the incorporation of von Neumann into the team, the design of the apparatus and its automatism were improved.[49] This led to the creation of the EDVAC (Electronic Discrete Variable Automatic Computer), the first direct successor to the ENIAC, and this was then followed by a whole string of computers designed according to von Neumann's architecture. Financed by the US Army, these two devices are usually taken to be the first electronic computers in the world, without overlooking the previously mentioned forerunners. For my part, I consider them the initial paradigm of techno-mathematics, following the notion of technological paradigm in a similar way to that of Dosi: "In broad analogy with the Kuhnian definition of a 'scientific paradigm,' we shall define a 'technological paradigm' as a 'model' and a 'pattern' of solution of *selected* technological problems based on *selected* principles derived from natural sciences and

49. One should remember that World War II also produced enormous advances in automatic control technologies that had been especially developed in the field of naval engineering. The German V-2 rocket and American B-29 bomber introduced this technology into the field of battle, with spectacular results.

on *selected* material technologies."[50]

In the development from the ENIAC to the EDVAC, the main innovation consisted of the fact that the program that arranged the carrying out of calculations was recorded in the same machine. In this way, the original notion of what today we call "software" began to be developed, given that the machine was not limited to just calculating but also processed information and then made calculations based on this.[51] The EDVAC therefore contributed another technological innovation in rendering a new action possible that, until then, had only been carried out by the human brain: processing information.[52] I would underscore the importance of this development, which could be considered a disruptive innovation in the Schumpeterian sense of the term. It was carried out by a research group that was working in a military technoscientific agency, the US Army's Ballistic Research Laboratory. In times of war, then, the profile of Schumpeterian entrepreneurs can change significantly.

I would conclude that, in the Age of Technoscience, one should pay attention to that area of entrepreneurship that emerged in public defense and security institutions. Some of the most important innovations of the twentieth century, including for example the Advanced Research Projects Agency Network (ARPANET) and the Internet, originated in such technoscientific institutions. Another thing is their later dissemination in civil society as a whole, a phase that was also important in the case of the ENIAC and the EDVAC. I would likewise conclude that innovation studies must be based on the idea of progress, distinguishing the various phases in each innovation process and studying them separately, and even, where appropriate, using different methodologies.

I will not explore the ENIAC project, which has been widely studied,[53] in any more detail. I would point out, above all, its huge

50. Dosi, "Technological Paradigms and Technological Trajectories," 152.

51. Breton, *Historia y crítica de la informática*, 93.

52. Von Neumann was aware of Alan Turing's work on symbolic computation and applied the neurophysiological theories of Walter Pitts and Warren Sturgis McCulloch so that the EDVAC would resemble the structure of the human brain.

53. See for example McCartney, *ENIAC: The Triumphs and Tragedies of the*

economic cost and extremely complex nature. It is highly improbable that any private entrepreneur would ever have been able to undertake such a wide-ranging project. Even in the case of a government, the wartime context was decisive for being able to develop missions like the Manhattan Project or the ENIAC, despite the enormous financing they required. From a technological perspective, the ENIAC implied a major advance because it computed the trajectory of missiles and shockwaves very quickly, thereby solving problems pertinent to military bombing. Let us return once more to the Schumpeterian definition of innovation: the ENIAC and the EDVAC radically transformed a way of doing something (mathematical calculations) and also facilitated new mathematical actions (calculating trajectories of long-range missiles, simulating explosions, processing information, and so on), which in turn gave rise to new military actions. As we will see in chapter 5, innovations, whether they involve a product or a process, are organizational, or are of another variety, transform the space of capacity of action.

Nor should one overlook the scientific knowledge that von Neumann contributed to the ENIAC project, both from the field of physics and mathematics as well as the emerging theories of computation, some of the latter linked to neurophysiology. Finally, let me make some brief notes about the diffusion process of these computational innovations. Once the war was over, important discrepancies emerged among the team members that had designed and built it. Eckert and Mauchly wanted to market the machine and indeed ended up founding UNIVAC. The US Army built new and more powerful prototypes exclusively for military use and kept this new area of research secret. As regards von Neumann, he used these computer prototypes for scientific research and spent his time giving lectures all over the world in order to publicize the innovation in line with the scientific ethos norm of making knowledge public. With the war over, the ENIAC continued to be of military and scientific use, but it also began to spread via the market. This example demonstrates that the diffusion phase can be developed in several ways and not just via markets. Fi-

---

World's First Computer.

nally, there was a problem with patents and uses that was first settled in court and later in tough commercial competition between UNIVAC and IBM, with the latter appointing von Neumann as a scientific advisor. The technological innovations that the ENIAC contributed and the resulting computers were applied in the military, scientific, and commercial fields, which gave rise to conflicts of interests and values among members of the initial team. It can thus be verified that one must analyze the dissemination of innovations in detail, given that this generates debates and problems that differ to those of the design and construction phase of a technological innovation. I would conclude that there are several potential models to analyze innovation processes as well as a plurality of valuation criteria. The ENIAC generated economic value, but also produced valuable applications in the scientific and of course military worlds, having emerged out of the latter in the first place. In short, innovations originate in many different places, and are not just generated by entrepreneurs.

The Evolution of Big Science

In 1962, Derek J. de Solla Price, a physicist and historian of science, gave the Pegram lectures at the Brookhaven National Laboratory, one of the most important sites in the United States for the peaceful application of nuclear energy. During the lectures, he introduces a quantitative methodology for the study of science: at the outset he asks, "Why should we not turn the tools of science on science itself? Why not measure and generalize, make hypotheses, and derive conclusions?"[54] In line with his training as a physicist, Solla Price was interested in the shape and size of science rather than focusing on contents, theories, and discoveries, as philosophers and historians of science had done since the beginning of the twentieth century. "Treating science as a measurable entity," he contends, "I shall attempt to develop a calculus of scientific manpower, literature, talent, and expenditure on a national and on an international scale."[55] Such lofty aspirations are currently termed *indicators of scientific development* and they are very important when

54. Solla Price, *Little Science, Big Science*, ed. 1962, Preface, i.
55. Solla Price, *Little Science, Big Science*, ibid., p. ii.

it comes to shaping scientific policies. This line of inquiry produced a new discipline, scientometrics, which really took off in the second half of the twentieth century and forms part of quantitative studies on science and technology. Indeed, quantitative studies on innovation originate in scientometrics and also use the same indicators.

The data Solla Price presents, all taken from the United States, allows him to argue that science had grown exponentially in size through the twentieth century. This growth affected the number of scientists, publications, new discoveries, and also the financing of scientific activity. To mention one example, somewhere between 80 and 90 percent of all scientists that had ever existed in history were alive during the 1960s. Such dizzy growth leads him to propose the hypothesis that science had entered into a new phase, Big Science or macro-science. He portrays this through two mathematical laws: the law of exponential growth and the law of saturation. The former contends that, "science grows at compound interest, multiplying by some fixed amount in equal periods of time."[56] He establishes a period of fifteen years in which the size of science doubles. The latter clarifies this exponential dimension and suggests the logistic curve as a model of growth, according to which exponential growth that doubles every fifteen years is just the beginning of a logistic curve that later levels off at its saturation point. At this point, growth could stagnate, in which case science would enter into a phase of "senility," or recover its exponential tempo, thereby resuming a new stage of accelerated growth.[57] Forty years later, there is no doubt whatsoever that the first of these hypotheses is incorrect. As regards the second, it depends on what we term "science," as I will demonstrate in the next section.

The data and mathematical models proposed by Solla Price were subsequently corrected and refined by experts in scientometrics, but to some extent are still valid. From a conceptual perspective, it is interesting that the law of exponential growth led him to make a distinction that is widely accepted among scientists today: that there is Little Science (that which existed between the

56. Ibid., 5.
57. Ibid., 23.

seventeenth and nineteenth centuries) and Big Science (that of the twentieth century). Both are distinguished by their growth rate – very slow in the former and very fast in the latter. For my part, I believe that Big Science implies an evolutionary leap or mutation in the history of science and that it heralds the emergence of the technosciences, but I also think both can exist side by side one another, complying with different types of knowledge.

The notion of macro-science (Big Science) had been suggested in a 1961 article by Alvin M. Weinberg, who had proposed an economic criterion to define it: for a project to be considered macro-scientific, he contends in the article, its undertaking must involve a significant portion of the gross domestic product (GDP) of a country.[58] According to Weinberg's criterion, the distinction between science and macro-science is above all one of the budget assigned.[59] Solla Price accepts this economic criterion but seeks to determine it more precisely and formalize it. In order to do so, he proposes a mathematical model that justifies the need to significantly increase spending on science. The conceptual distinction that he introduces contends that scientific research had embarked upon a new historical phase and because of this it was necessary to rethink the issue of funding. He also leaves the way open to differentiate macro-science from science not just according to its size, but also on the basis of qualitative and cultural criteria.

The same question was reexamined in a symposium organized by Stanford University in 1988, whose proceedings were edited by two historians of science, Peter Galison and Bruce Hevly. For Hevly, "'Big science is not simply science carried out with big or expensive instruments."[60] Big budgets and large instruments are indicators of change, yet according to Hevly, macro-science was characterized from its very beginning by the following changes:

58. Weinberg, "Impact of Large-scale Science on the United States," 164. See also Weinberg, *Reflections on Big Science*.

59. This criterion used to define macro-science subsequently became a standard measure. In the *World Inventory of 'Big Science' Research Instruments and Facilities*, published by the US Congress in 1986, macro-science is mentioned to allude to large scientific apparatuses constructed after 1920 and costing a minimum of twenty-five million dollars in constant dollars at the 1984 rate.

60. Hevly, "Reflections on Big Science and Big History," 356.

the concentration of resources at a few elite research institutions, specialized workforces in laboratories, and the development of projects that were significant from a social and political point of view and that contributed to increasing military power, industrial potential, and the health or prestige of a country.[61]

During that same symposium the origins of macro-science were also debated as well as its later evolution. Hevly highlights other particular features of macro-science that became increasingly more evident with time: namely, that the relationship between science and technology took on new forms that in turn influenced the nature of both of them, and that macro-science requires the interaction of scientists, engineers, and the military. Galison confirms this latter point, in emphatically stating, "It is manifestly impossible to examine Big Science without addressing the science of war."[62]

Other authors also present their definitions of macro-science in the published proceedings of the symposium. Robert W. Smith, for example, refers to the politicization and bureaucratization of science management, as well as an increase in the risks taken and the loss of autonomy on the part of the scientific community.[63] In his introduction to the work, Galison contends that macro-science has many faces, with the result that any examination of it must necessarily be difficult and complex. And all the authors in the work point out that the discontinuity between Little and Big Science is in part fictitious, although this does not hamper the convenience of maintaining the distinction. W.K.H. Panofsky, one of the major figures behind Big Science at Stanford University (in the Mark III linear accelerator project, for example), refers to a difference of scale between forms of science.[64] In doing so, he invokes once more the initial ideas of Solla Price, emphasizing that the transition from science to macro-science was evolutionary, not revolutionary.

In the same work, authors also address the emergence of macro-science in Europe (the CERN) and Japan (the science city of

61. Ibid., 356–57.
62. Galison, "Introduction: The Many Faces of Big Science," 13.
63. Smith, "The Biggest Kind of Big Science," 186.
64. Panofsky, "SLAC and Big Science," 145.

Tsukuba), showing that there are significant cultural differences in its development according to both country and discipline. At the end of the day, both the participants at the Stanford University symposium and other authors who have addressed this subject coincide in their use of the term Big Science to allude to a new stage in the development of science, but differ when it comes to attempting to specify what this is exactly and define it more precisely. Most studies would subscribe to the general observation offered by José Manuel Sánchez Ron, according to which, "Big Science is a research procedure characteristic of our century,"[65] referring to the twentieth century. Yet there are discrepancies over the exact definition. For my part, I have shown elsewhere that through the twentieth century the growth of science changed not just in terms of size and rhythm but also in a much more profound way: namely, in terms of the structure of technoscientific activity.[66] Weinberg's economic characterization and Solla Price's quantitative models are indicators of this change but they only address certain aspects of this new form of science. The emergence of macro-science implied a huge change in scientific practice from which many other changes – some of them of great magnitude – were subsequently derived. For that reason, I subscribe to the thesis that through the twentieth century there was a profound revolution in science and technology: a *technoscientific revolution*. The scientific revolutions that Thomas S. Kuhn studies (of Copernicus, Galileo, Newton, Lyell, Lavoisier, and Einstein, and in quantum mechanics and so on) transformed above all else the structure of scientific knowledge. The twentieth-century technoscientific revolution, however, was based on a radical change in the structure of scientific activity. These structural changes brought with them a growth in the size of science, but also modified scientific objectives, relations within scientific communities, ways of organizing research, and the evaluation criteria of the results. In particular, this transformation led to a symbiosis between science and technology. I also contend that the massive introduction of private capital and the introduction of the ICTs into scientific research, both of which were features of

---

65. Sánchez Ron, *Falsos mitos*, 36.
66. Echeverría, *La revolución tecnocientífica*, chap. 2.

this transformation from the 1980s on, entailed a new qualitative change that was especially evident in the final quarter of the twentieth century: macro-science, a transitional phase toward technoscience, properly speaking. Macro-science emerged especially in the 1980s, although it had existed previously in certain areas of knowledge. In my opinion, the great change experienced by science in the twentieth century is best examined when we speak in terms of technoscience rather than macro-science. For this reason, I argue that macro-science is the initial form of technoscience.

Solla Price locates the beginning of macro-science during the World War II era Galison, Sánchez Ron, and Robert Seidel, among others, point out the historical importance of Ernest Lawrence's design and construction of the first cyclotron at UC Berkeley in 1932. Consequently, they tend to date the beginnings of Big Science to the decade prior to World War II.[67] Lawrence's project is indeed a good example of a macro-scientific proposal, although one should bear in mind that the large cyclotrons and particle accelerators were built in the 1940s and beyond. For me, debate over a specific moment at which macro-science emerged is futile. This is not a discovery that can be attributed to one person in particular, nor is there a specific date in time or a precise geographical location by which to situate it.[68] Instead, this is a *progressive change*

67. Galison, "Introduction: The Many Faces of Big Science," 3; Sánchez Ron, *Falsos mitos*, 38; Seidel, "The Origins of the Lawrence Berkeley Laboratory," 21–45. Lawrence's 1932 cyclotron measured 30 centimeters (27 inches) in diameter and 1 million volts in power. In 1937, a 16-million-volt cyclotron was constructed at Berkeley. The size and power of cyclotrons increased enormously in the 1940s and eventually arrived at the celebrated Superconductor project, ultimately aborted by the US Congress in 1993, which would have had a 60-mile-long tunnel, twenty times more energy than anything existent at the time, and an estimated cost of $8 billion dollars. Other authors, such as Michael Dennis, demonstrate persuasively that creating temporal boundaries between Little Science and Big Science, as well as locating a precise historical origin of the latter, implies a kind of belief in myths. See Dennis, "A Change of State." Here, Dennis argues that macro-science already existed at two US universities in the 1930s.

68. One should recall that both Robert K. Merton and Kuhn underscore the fact that scientific discoveries are frequently produced simultaneously and independently of one another. This is attested to by the numerous historical controversies that exist as to the order in which the discoveries were made.

in the structure of scientific activity that requires a sufficiently long period of time in which to emerge, develop, and be consolidated. The technoscientific revolution was not carried out by one person or by one research center. Nor was it an epistemological, methodological, or theoretical change like that of the seventeenth-century scientific revolution. It was a radical transformation in research activity that took place in several research centers at the same time, although in some of these the ideas associated with it took shape more quickly and more clearly than in others. Moreover, it did not just take place in laboratories and research centers but in other places as well (science policy offices, companies, foundations, strategic studies centers, and so on). Meanwhile, the emergence of technoscience did not just affect research but also the management, application, evaluation, development, and diffusion of science — in other words, science as a whole. The technoscientific revolution has been a prolonged and complex process that is still unfolding insofar as technoscience continues to expand within different countries. In each of these, it has taken on different forms in the same way as it changes according to different disciplines. Having said that, there are still a series of common features that allow us to characterize the new structure of scientific-technological practice, as we will see in the next section.

Let us proceed step-by-step, casting our minds back to the first form of technoscience or Big Science. Macro-science entailed a new scientific-technological system, and for that reason there were numerous and important changes in scientific practice. Some American universities and research centers (MIT, UC Berkeley, and so on) had already begun to go in this direction in the 1930s. One can, then, speak about several examples of macro-science prior to World War II such as the MIT Radiation Laboratory, the Klystron Laboratory at Stanford, and the Berkeley Radiation Laboratory. Nevertheless, the main surge in research centers like this took place during the war and its immediate aftermath. What these centers hinted at in the 1930s demonstrated their enormous effectiveness in World War II. The Bush report added a theoretical dimension to this change, thereby facilitating the implementation of a new science and technology system in the United States following World War II. This system was consolidated in the postwar era and spread farther afield during the 1950s, first in the United States itself and later in other countries, with their corresponding variations and specificities.

Therefore, I situate the beginning of macro-science in the World War II era, highlighting the fact that it emerged in the United States and in the field of militarized physics and mathematics, as the four great projects that might be considered initially canonical of Big Science demonstrate: the Berkeley Radiation Laboratory, the MIT Radiation Laboratory, the ENIAC project at the University of Pennsylvania's Moore School of Electrical Engineering, and especially the Manhattan Project at Los Alamos, New Mexico, an authentic paradigm of macro-science.[69] Other countries such as Germany and Great Britain developed similar projects both during and even before the war.[70] Yet in the postwar era, they lacked the sufficient funding and political support to promote technoscience. Those countries instead prioritized reconstruction of their devastated societies over the creation of a national system of science and technology that would have made them world leaders during this time, as was the case in the United States. In the American case, the emerging technoscience contributed decisively to the US military victory during the war. But the key decision in the longer term was that of encouraging a new organizational structure for macro-science during the postwar era, and dedicating

69. The MIT Radiation Laboratory was founded in 1940 and quickly came to specialize in the radar project, which was of major scientific – but also technological and military – importance. Following the war, radars were also applied widely in aviation and civil navigation. Following this line of research, Bell Laboratories invented the first transistor, another major example of technoscientific innovation, in 1947. As regards the ENIAC project, it led to the construction of the first multiple-function computer, whose design was overseen by von Neumann but which would have been impossible without close collaboration among physicists, mathematicians, and engineers. In the same way as the Manhattan Project, it was financed by the US Army and its results remained top secret during the war. See Sánchez Ron, *Falsos mitos*, 43–52; Seidel, "The Origins of the Lawrence Berkeley Laboratory," 21–45.

70. Cryptology would be another example of macro-science during this era, with competing British, German, French, and American projects being established.

significant public funding to this end. In this way, certain military institutions became significant agents in scientific and technological research.[71] Once the war was over, the only country in sufficient economic, political, and military condition to undertake Big Science was the United States. As a result, macro-science emerged in the four previously mentioned centers (as well as others), but the new structuring of scientific activity was not carried out until 1945, specifically following official approval of the Bush report and its gradual establishment in practice.

Some years later the Soviet Union became a nuclear power, developed its own science and technology system, and took the initiative in space exploration with the launching of the first artificial satellite, Sputnik, in 1957. Therefore Big Science was also developed in the Soviet Union, although later than in the United States. The scientific and technological competition between the two superpowers was one of the most significant features of the Cold War and closely linked to their military, political, industrial, and ideological rivalry. The struggle to control space between the United States and the Soviet Union in the 1950s and 1960s is an excellent example of macro-science, as is the strong rivalry between the Brookhaven National Laboratory in the United States and the CERN in Europe (established in 1952) in the field of small-particle physics. There are more and more examples of macro-science from the 1950s on, and macro-science began to be embraced by other industrially developed countries from the 1980s on. In the latter case, though, there were significant differences as regards the initial phase of this development in comparison to the American case, with governments providing the principal motivation and financial stimulus for large-scale projects. Greater involvement on the part of private companies and the computerization of technoscientific activity were, in turn, more features of macro-science's second phase, in which what one might more properly say technoscience took shape. For these reasons, I would argue that macro-science was the precursor of technoscience, and that both differed from modern science as regards not just their quantitative

71. In 1955, 80 percent of federal government investment in R&D was channeled through the Department of Defense.

but also their qualitative criteria. And a key moment in this change was the drawing up of the Bush report, which is I why I pay special attention to it here.[72]

The Concept of Technoscience

"Technoscience" is a controversial term. Scientists involved in basic or fundamental scientific research typically look at it suspiciously because it appears to prioritize technology and applied research. Philologists consider it a barbarism because it combines two lexical roots of Greek and Latin origins. Many philosophers of science prefer to continue establishing clear lines of demarcation between science and technology, fearing that in discussing technoscience the specific nature of science might disappear, devoured by technology. Others, in contrast, contend that technoscience has been a reality since the nineteenth century and even earlier.[73] For me, this expression helps to comprehend and interpret the profound changes in scientific and technological activity throughout the twentieth century.

The term "technoscience" was coined by Bruno Letour in *Science in Action* (1983), with the aim of avoiding "endless 'science and technology'" references.[74] Latour asks the question, "Who is really doing science, after all?"[75] and attempts to show that science is not just done by scientists. With this in mind, he criticizes the internal/external division that is prevalent when it comes to thinking about science.[76] According to this division:

- Scientists, and especially research scientists, are active within science; outside science, there are politicians, busi-

72. On the Bush report and the legislative initiatives that preceded it, see Smith, *American Science Policy since World War II*.

73. See, for example, Moya, *Crítica de la razón tecnocientífica*, 25, in which the author observes, "By technoscience we understand that kind of reason that emerged in the West during modernity and whose main aims were the search for truth and efficiency. Agazzi, too, although in a different way, argues this point, at least in *La techno-science et l'identité de l'homme contemporain*.

74. Latour, *Science in Action*, 29n.

75. Ibid., 157.

76. Ibid., 157–59.

nessmen, teachers, lawyers, and so on.

- Science is done in experimental laboratories and disseminated at conferences and in scholarly journals, in which the scientific community debates and comes to agreement on ideas that originate in laboratories and are considered acceptable and valid, even though still only conjectural.
- Once elaborated, this knowledge is diffused throughout society and applied in order to solve practical problems. During this phase, science generates technology, which is identified with applied science. This is the moment when science comes into contact with the external world: society. Until then, science has been internal.

According to Latour, this "diffusion model"[77] of science is inadequate. And the supposed boundary between inside and outside is fictitious.[78] Latour is correct in his criticism of this model and especially in refuting the identification between technology and applied science. Some philosophers of technology have also argued that this identification is erroneous.[79] This is basically due to the fact that there are multiple historical examples of technology having its own technological paradigms and trajectories, to use the terminology of Nelson, Winter, and Dosi.[80] For my part, I believe that science and technology were separate until the emergence of technoscience, although very close links were established between the two during the Industrial Revolution. Another thinker who systematically uses the term "technoscience" is Gilbert Hottois.[81]

I am, throughout this chapter, attempting to detail and elucidate the concept of technoscience. Before proceeding with this aim, it is worth explaining the conceptual framework in which I will offer my proposals and clarifications.

First, I do not seek to define the notion of technoscience. I

---

77. Ibid., 133.
78. Ibid., 162. Friedrich Rapp makes identification between technology and applied science in his *Analytical Philosophy of Technology*, as do other authors such as Mario Bunge.
79. See for example Mitcham, *Thinking through Technology*.
80. Nelson and Winter, "In Search of a Useful Theory of Innovation"; Dosi, "Technological Paradigms and Technological Trajectories."
81. See for example Hottois, *La philosophie des technosciences*.

will instead limit myself to outlining a series of distinctive features common to science, technology, and technoscience.

Second, these distinctions will be measured in terms of degrees of difference rather than necessarily serving to demarcate specifically the boundaries among them. I am not trying to establish borders between science and technoscience, given that the latter is a particular type of science. That said, the different levels associated with each that I will suggest below will permit me to distinguish one from the other with sufficient clarity, in most cases by the extent to which both satisfy these levels.

Third, I will not resort to the method of specific difference, but rather to that of features and different levels. I will not, then, concern myself with the difference between science and technoscience, supposing that exists, but rather with the set of distinguishing features that may be discerned between the two. There are of course several differences, although none of these could be considered decisive or "essential." In order to classify an activity as technoscientific, it should satisfy many and not just one of these distinguishing features. Once the concept of technoscience has been elucidated via this methodology, we will have at our disposal a clearer and more precise concept, even though I recognize that this clarification can of course be improved on.

Fourth, my perspective is not reductionist. Although technoscience may have emerged throughout the twentieth century, human beings continue to (and will continue to) develop technical, technological, and scientific activities. In all likelihood, technoscience will undergo major development during the twenty-first century. However, this does not imply that science and technology will disappear. Arts, sciences, technique, and technologies exist and will continue to exist. Moreover, there are also technosciences. It is a question here of highlighting this latter and increasingly important variety of knowledge without disregarding the rest.

Fifth, the emergence of technoscience could be considered as a new kind of revolution, although not in Kuhn's sense of the term. For that reason, it will be necessary to define exactly what might be meant by the term "technoscientific revolution."

Sixth, I will discuss "action" and "technoscientific activity" rather than scientific knowledge and technological artifacts. This is the main change I propose when it comes to examining the technoscientific revolution. Twentieth-century philosophy of science focused – with very few exceptions, such as Ian Hacking and Joseph Rouse[82] – on analyzing, reconstructing, and where applicable, justifying and bolstering scientific knowledge. Meanwhile, the sociology of scientific knowledge (the strong program or strong sociology, the empirical program of relativism or EPOR, social constructivism, and so on) was concerned almost exclusively with the social construction of scientific knowledge. However, there was an important praxeological shift in the late twentieth century[83] and an emerging interest in technology on the part of sociology. In recent years some historians of science, including Allan Franklin, Buchwald, and Peter Galison,[84] have also begun to take an interest in scientific practice and above all in experimentation. In my view, studies of scientific practice require a theory of technoscientific action; it is not enough to just have an epistemology and a methodology. They must draw up a praxeology or, in other words, a theory of technoscientific practice, a task in which several authors are involved.[85] Technoscience emerged out of a profound change in the structure of scientific practice, not through an epistemological or methodological revolution.

Seventh, within that praxeology one must develop an axiology of technoscience. If we conceive of science and technology as activities, then it is easier to compare them and determine their differences. I share Evandro Agazzi's view that human actions are guided by values.[86] Furthermore, my approach to the axiology of technoscience is grounded in the notion of value systems in which the meaning and application of each value depends on all the others in the system, in such a way that it varies over time and accord-

82. See Hacking, *Representing and Intervening*; Rouse, *Engaging Science*.
83. See Pickering, ed., *Science as Practice and Culture*; Pickering, *The Mangle of Practice*.
84. Franklin, *The Neglect of Experiment*; Buchwald, ed., *Scientific Practice*; Galison, *How Experiments End*.
85. See for example the recent work by Martínez, Huang, and Guillamin, eds., *Historia, Prácticas y Estilos en la Filosofía de la Ciencia*.
86. See Agazzi, *Right, Wrong and Science*, 63–72, in which he defines technoscience as an ideology.

ing to different contexts, as opposed to a static essentialist conception of values. Thus, there are significant differences among the value systems that guide technical, scientific, technological, and technoscientific actions. I will, likewise, say the same thing about innovation values in chapter 5. The following thoughts anticipate the philosophy of innovation I seek to propose in this book.

The Distinction between Sciences and Technosciences

Almost all experts in science and technology studies accept that science changed drastically through the twentieth century, although they differ when it comes to conceptualizing this change. J. M. Ziman distinguishes between "academic" and "post-academic" science with the aim of describing a "radical, irreversible, worldwide transformation in the way that science is organized, managed and performed."[87] In his opinion, "science is being redefined at every level, and in relation to other segments of society. This is what I mean when I say . . . that academic science is giving way to post-academic science."[88] Ziman's ideas were very influential in science policy during the 1980s, especially in English-speaking countries, partly due to his position as director of the Science Policy Support Group in Great Britain and partly because of his theoretical writings on the social dimension of science and science policy. As noted, Latour proposes the term "technoscience" in order to underscore the close connection between science and technology in the twentieth century, although he later uses it as a technical concept with a precise meaning linked to the close association between scientists and their instruments. According to his actor-network theory, there is non-human technological agency in research activity and, as a consequence, clear hybridization between science and technology.[89] Silvio Funtowicz and Jerome Ravetz, meanwhile, talk in terms of "post-normal science," a term that addresses the problems stemming from Kuhn's disciplinary matrixes.[90] They have also argued that, in such cases, scientists act in conditions of

87. Ziman, *Real Science*, 67.
88. Ibid.
89. See Latour, *Reassembling the Social*.
90. See Funtowicz and Ravetz, "Science for the Post-Normal Age" and *La ciencia posnormal*.

uncertainty, so that science is not subject to any form of determinism. In a prior work, Ravetz also distinguishes between "academic" and "industrialized" science.[91] Elsewhere, Helga Nowotny, Peter Scott, and Michael T. Gibbons (among other authors) argue the case for a new form of knowledge production – Mode-2 knowledge production – that is multidisciplinary, heterogeneous, and non-hierarchical.[92] Herbert Simon, on the other hand, refers to design science.[93] And Henry Etzkowitz proposes a triple-helix model made up of universities, industry, and government, which both he and Loet Leydesdorff develop.[94]

These authors and others who study science and technology coincide in arguing that, since the emergence of Big Science, science has changed radically, especially as a result of the development of the ICTs. The term "e-science" is another way of designating that transformation in contemporary science and is related to the emergence of an information society and knowledge economy. Nobody, then, denies the fact that science has changed or that it is closely linked to technology, and in particular to the ICTs.

This is one of the reasons I prefer to use the term "technoscience." However, in contrast to other authors (including Latour), I do not use it as an all-inclusive term that encompasses everything. The conceptual framework I propose might be summed up in the following way:

1. From a philosophical point of view, the main difference between science and technoscience is down to knowledge: for science, the quest for knowledge is an end in itself, whereas for technoscience it is a means to an end. Technoscientific companies and agencies are interested in the quest for knowledge, as well as its validation, but their objectives go further than just this. In RDI systems, priority areas of research are those that generate technological developments and, in the final analysis, innovations. There

91. Ravetz, *Scientific Knowledge and Its Social Problems*, 5.
92. Nowotny, Scott, and Gibbons, *Re-thinking Science*.
93. Simon, *The Sciences of the Artificial*.
94. Etzkowitz and Leydesdorff, eds., *Universities and the Global Knowledge Economy*.

is science when the quest for knowledge continues to be the main objective. There is technoscience when scientific knowledge becomes a means to generate new technologies and innovations.

2. Science attempts to explain the world through phenomena and, where applicable, predict them. Technoscience, on the other hand, attempts to change the world, and not just the natural world but also the social world.
3. The sciences still exist; *not everything is technoscience*. Latour, Hottois, and other authors tend to suggest that technoscience has absorbed science. I do not believe this to be true and think we have to distinguish carefully between the sciences and technosciences, including the distinction between social sciences and social technosciences. It is important to analyze when a technoscientific discipline emerges and how it does so. We will see below that the social technosciences use complex non-determinist computational models that are very different to traditional mathematical models.
4. The transformation mentioned does not just affect knowledge but also, and mainly, scientific practice. On this point I differ from Gibbons, Nowotny, and all those who defend the notion of Mode-2 knowledge. There has not just been a change in the way scientific knowledge is produced but also in the way it is presented, evaluated, distributed (published), stored, and used. Examples abound. The evaluations of impact factor indicators such as that of Thomson-Reuters are a good example of social technoscience, which modifies scientific practice and its evaluative criteria. One could say the same for the online teaching of science (through virtual campuses), online publications (ArXiv, Scopus, electronic journals, and so on), the procedures for evaluating research projects, and digital repositories (for example through the 2004 Berlin Declaration on open access to knowledge in the sciences and humanities). The ICTs now permeate all areas of technoscientific activity, not just knowledge production. For that reason, I con-

tend that during the second half of the twentieth century there was a technoscientific revolution, as a result of which national RDI systems emerged whose ultimate objectives are innovation, to which research is subordinate.

5. From a sociological perspective, scientific communities undertake academic science, as Merton demonstrates. Kuhn even closely correlates the notions of scientific community and paradigm. Technoscience, however, is done by technoscientific companies and agencies, and they define technoscientific agendas or what should be done. In the knowledge economy, scientists become "knowledge workers" and lose a large part of their autonomy in the process. Technoscientific practice necessarily includes internal conflicts because of the structure of its agency, which is very different to that of science.
6. As a result, in the twenty-first century it is not enough to just do the philosophy of science or the history of science. Besides these, one must also do the philosophy and history of technoscience, including a philosophy of innovation, which is not the same as a philosophy of scientific research. Neither epistemology nor methodology is sufficient when it comes to analyzing technoscience. What is needed, above all, is a *philosophy of technoscientific practice* that includes a theory of scientific and technoscientific action and not just a theory of scientific knowledge. The idea of a technoscientific agenda, which I will now discuss, is in my opinion crucial.

There is also fairly widespread consensus when it comes to locating this great change in science spatially and temporally, specifically to the United States during World War II. An important role is attributed to Bush's report *Science, the Endless Frontier* (1945) as well as to the linear development model, RDI. According to this model, scientific research is the principal source of technological developments and innovations and it should, therefore, be generously funded. This model was established in almost all developed countries during the last third of the twentieth century, although in the early twentieth century it experienced a crisis.

As regards the main topics addressed in this work, I would argue that an important structural transformation in scientific practice came about after 1980 insofar as innovation became the main purpose of R&D. This brought with it major changes to the structure of these systems, as well as to science and technology policies. In order to be leaders in their fields, technoscientific groups must help to generate innovations and not just knowledge as reflected in high-impact-factor publications. In order to do so, they must create companies or at least be closely connected to them. Scientific research in itself is not enough for technoscience. This second change could be expressed by saying that scientific communities have ceased to be the subject agent of science, being substituted in this role by both public and private technoscientific companies whose strategies are guided by the imperative to innovate. Managing scientific research in a businesslike way and declaring innovation to be the ultimate goal are two distinctive features of current technoscience.

Technoscientific Agency

There are several features that allow us to distinguish the sciences from the technosciences. In this section I will look at one of these in detail that is related to the emergence of a new "subject of science," the technoscientific firm. Let us examine this briefly.

Andrew Pickering proposes the notion of scientific agency.[95] I agree with this, although I reinterpret the idea. I believe that notions like technoscientific agency and technoscientific agenda are basic as regards the philosophy of technoscientific practice, but also as regards technoscience policy. One should note that they are comparable to the idea of scientific theory for dominant science, which was focused on knowledge, in the twentieth century.

I take the term "technoscientific agency" to mean an organized plurality of individual, institutional, and business agents that promote and undertake technoscientific projects and activities. It therefore not only includes scientists, engineers, and technicians but also other kinds of professionals. I am thinking, here, about

95. See Pickering, ed., *Science as Practice and Culture*; Pickering, *The Mangle of Practice*.

government agencies (such as the NSF, NASA, ESF, CERN, and so on), and also about numerous private technoscientific firms, always on the understanding that they do scientific research oriented toward innovation.

A laboratory or a research team are scientific agents. But technoscientific agents are larger and more complex: they could be, for example, an R&D company like Monsanto, a national research agency such as the NSF, or a major institution made up of representatives of several countries such as the CERN in Europe. Although there are always individual scientists working in these agencies, the actions taken there are often in the name of the agencies themselves: for example, sending a spacecraft to Mars, researching the Higgs boson, and so on. Scientists work in technoscientific agencies and do science there; in other words, they follow the guidelines of modern scientific practice: they observe, experiment, contrast hypotheses, communicate the results to their colleagues, publish articles in specialized journals, get recognition, gain prestige, and so forth. However, technoscientific agencies and companies do many more things, besides research. Other kinds of agents are superimposed onto the scientific community who take charge of managing research infrastructures, administering human and economic resources, obtaining patents, getting funding through projects and contracts, publicizing the firm itself, and so on. Technoscientific agencies and companies use costly research infrastructures that must be maintained and renewed; patent those results that are interesting financially; establish deals with politicians, the military, and other firms; take charge of managing material and human resources; defend their own interests through legal departments; carry out marketing with the results of their research; reinvest any profits they may make; and so on. Such actions and many more one could mention are not scientific but business and management actions. This is one of the differences between scientific and technoscientific practice. Scientists are just one (and often not even the most important) component of a technoscientific agency. Scientific knowledge is a necessary condition of technoscientific activity but is not in any way sufficient in itself for there to be technoscience. All the remaining components (investment, funding, infrastructures, technical staff, management teams, legal departments, deals with lobbies with access to decision-making authority in science policy, marketing and human resources departments, and so on) are also necessary conditions for technoscience. In a word, technoscientific agencies and companies make up much more complex bodies than the scientific communities theorized by Merton and Kuhn.

I maintain the philosophical hypothesis that the transition from science to technoscience has changed the subject of science, transforming it into a diverse subject. A minimally developed and structured technoscientific firm must have at its disposal, as well as scientists, engineers, and laboratory technicians, other professionals: managers, advisors, experts in marketing and work organization, lawyers, contacts in political and military circles, investors, financial bodies backing them, and so on. A technoscientific agency is never formed by one individual alone and nor can it be reduced to just a research group. Included within technoscientific companies, and as indispensable components therein, is a great diversity of experts. They all carry out necessary functions for the agency, although later prestigious scientists may appear as the spokespeople when it comes to publicly presenting the results obtained (in the event the results are public). If, on the other hand, the results are patented, the company spokesperson is not a prestigious scientist but a legal expert specializing in issues of intellectual property. With the coming of national science and technology systems there is not just a new form of knowledge, technoscience, but an altered internal structure of scientific practice. The internal dimension of technoscience differs greatly from that of science, in case we still wish to preserve the interior/exterior distinction. There has been much debate within the philosophy of science about the objective nature of scientific knowledge or, to put it in Karl Popper's terms, epistemology without a knowing subject. After a learning process, any human can accept scientific knowledge and make it his or her own. The minds of individual scientists, the "men of science," were the great deposits of knowledge, apart from journals, libraries, and the printed material communicated to the rest of the scientific community. In the case of technoscience, however, complex heterogeneous teams of people are required, as well as different means and instruments. What is most important is appropriating ways of carrying out technoscientific actions. Given that it so happens that there are extremely varied kinds of actions, the subject of technoscience is plural. Or perhaps it is better to not even speak about a subject but, rather, an agent, actor, doer, or an active individual.[96] Whatever the case, this is always a plural form of agency because it requires the cooperation of different kinds of experts and numerous artifacts for a technoscientific action to produce acceptable results. From the individual subject of modern science (the genius), one thus moves on to a research team with a business, administrative, political, and legal support structure. For the results of scientific research to be fully acceptable, it is not enough to just have epistemic contributions. Moreover, it is necessary for scientific knowledge to generate technological developments and innovations, in such a way that this knowledge is transferred to companies and institutions. As a consequence, the very notion of acceptability changes. It is no longer enough for a scientific hypothesis to be verified empirically. The important thing is that the concomitant technological developments, too, have been verified technically and, above all, that any potential derived innovations have been validated according to two very specific reference points: markets and societies. These are both, in the final analysis, the points at which technoscientific proposals are accepted or rejected.

This profound change can be summed up in terms of an evolution from R&D systems to RDI systems in the late twentieth century. The third letter in the acronym came later than the other two, but today it is the principal one of the three, especially for knowledge companies, as in the case of technoscientific businesses. The ultimate aim of technoscientific activity is not limited to the advance of knowledge alone, nor even to any technological developments produced. The ultimate aim is innovation, which implies

96. The actor-network theory of Latour, Michel Callon, and John Law contributes a very interesting additional nuance to the notion of the plural subject: a lot of technoscientific actions are carried out by machines, artifacts, robots, and the like. For that reason, not just human actors or agents do technoscience but also artificial agents. See Latour, *Reassembling the Social.*

the transformation of knowledge advancement into innovations that might improve the competitiveness of technoscientific firms. Put succinctly, science and technology are subordinate to innovation, for all that scientists preserve their autonomy and just concern themselves with knowledge in their laboratories and scholarly journals.

By contrast, innovative proposals are examined in markets and societies rather than in laboratories, where at most there can only be discoveries, attestations, and inventions. Moreover, scientists and engineers have their own evaluation criteria regarding the results of their research. Peer evaluation, high-impact academic journals, the Nobel Prizes, and so on still exist, but for a technoscientific proposal to be accepted other criteria are applied, such as the investment undertaken, research costs, possible returns, potential to generate innovations, comparison with other countries or competitor firms, and so on. Indicators of technoscientific activity are not limited to impact in the corresponding scientific community. The conditions in which national science and technology systems are measured have changed dramatically in the last twenty-five years, and as a result, scientific agencies have evolved and adopted a new structure.

In the twentieth century, the philosophy of science focused on scientific knowledge (theories, facts, conjectures, laws, and so forth), paying little attention to scientific practice. It attempted to justify scientific *episteme* through logic and reason (knowledge as a result of cause, verified, corroborated empirically, predictive, explicative, and so on) and even ground it, if on a logical base, in mathematics or physics. In the 1970s and 1980s, the sociology of science also focused on scientific knowledge. The philosophy of scientific practice, however, must analyze and, where appropriate, justify or critique technoscientific practice rationally. One way of doing so consists of turning to the value systems that guide such practice and estimate the results, at the same time evaluating each specific action, whether positively or negatively. Axiology is a good means of developing praxeology. This praxeology, however, is more extensive and dates back to Hacking's *Representing and Intervening* (1983). In his opinion, scientists do not just produce representations of the world because in order to do so they actively intervene in it. Facts, observations, measurements, calculations, demonstrations, experiments, conjectures, verifications, and refutations are the results of diverse scientific *tasks*: observing, measuring, calculating, demonstrating, experimenting, conjecturing, refuting, and so on. Science does not produce epistemic results without previous actions. The technical concept of scientific agency (Pickering) captures this shift very well, in contrast to the philosophy of science that focused on scientific theories alone. Any theory of scientific knowledge that lacks a prior theory of scientific practice is, then, philosophically insufficient. And for some years now, several authors have tried to clarify the basic concepts that underlie scientific practice. This task is undertaken by corroborating conceptual proposals with examples and case studies. Such methodology is canonical in science, technology, and society (STS) studies, as well as in certain currents within the sociology and history of science.

Recent contributions, some of them by Latin American authors,[97] have made important contributions to the philosophy of scientific practice. However, there is still much to do when it comes to examining its structure. This requires the contributions of scientists and experts in science and technology studies, with the aim of being able to check philosophical arguments against other analytical perspectives.

The Second Phase of Technoscience

Big Science experienced a profound crisis in the late 1960s owing in large measure to the failure of the United States in the Vietnam War. In contrast to the experiences of World War II, the Korean War, and the Cold War, technomilitary innovations were insufficient in Vietnam and the US military was not victorious. As a result, the linear model, which to that time had been the inspiration behind R&D, faced a crisis and military funding of basic scientific research declined sharply during the period 1965–1975.[98] It was a similar story with private funding, which fell by 36 percent in the

97. For example, Martínez, Huang, and Guillamin, eds., *Historia, Prácticas y Estilos en la Filosofía de la Ciencia*.

98. It fell by 25 percent in 1971 alone, and in the ten-year period between 1965 and 1975 it was reduced by half.

period 1966–1972. The scientific communities experienced this decline as a real crisis and even spoke about the rise of an irrational antiscientific movement against them.[99] Many universities closed down or reconverted research centers with defense connections.

The situation began to change during Gerald Ford's presidency (1974–1977) but especially with the two-term Ronald Reagan administration (1981–1989). In the 1980s, a new social contract for science was established that marked the starting point for the second phase of technoscience. From a budgetary point of view, there was a rapid increase in private funding for R&D thanks to a liberalization of patent laws and a new fiscal policy that allowed for up to 25 percent of private investment in R&D to be tax deductible. The priority was now technological innovation, with private initiative driving the system. This was still termed R&D, although it began to be thought of as a national RDI system. The US government still funded basic scientific research, but the principal aim of its science policy consisted of making private companies assume the main responsibility for this funding. In order to do this, the scope of what could be patented was extended greatly. Consequently, and not without prior debate, certain forms of knowledge in molecular biology (gene sequencing, for example) and computer sciences (software programs, different computer applications, and so on) began to be patented. This in turn brought a lot of private investment into the R&D sector, which had until that time been dominated by governmental technoscientific agencies. This new policy also entailed a radical change in the scientific research framework. From the 1980s on, private funding overtook its public counterpart and has continued to grow ever since, reaching over 70 percent of all investment in R&D in the United States. There was a similar process in Europe, although it came later and only in certain countries such as Germany. A new R&D phase was thus initiated in which technoscience, properly speaking, took shape. This was not centered on great scientific research

99. For example, on the part of Simon Ramo, who headed one of the two NSF committees in the late 1970s. He was strongly supported at the time by Vice President Nelson Rockefeller and the Ford administration. Indeed, it was Rockefeller who instigated the pro-technology shift that culminated during the Reagan administration.

projects (although some macro-research programs existed) but instead produced innovations. And these were sometimes not even done in a scientific laboratory but in other spaces such as the mythical garages of Silicon Valley. Some of these technological innovations – such as the ICTs and genetic engineering – gave rise to radical transformations and, in spreading socially, led to disruptive innovations. From a funding point of view, this phase was characterized by the primacy of the private over the public sector thanks to venture capital finance institutions. This change brought with it concomitant transformations, of which I will discuss just some here. In general terms, it implied a profound restructuring of the American science and technology system or, if one prefers, an organizational innovation by which private initiative and entrepreneurs assumed the leadership of R&D systems and transformed them into RDI systems, although always with the support of public funding.

For example, the stock market began to take an interest in investing in science and technology. In 1983, companies like Merrill Lynch and J.P. Morgan advised their customers to invest in R&D firms. In contrast to the funding of macro-science, which was predominantly governmental and military, technoscience found new ways to achieve funding, besides large corporations and government agencies. Small R&D companies proliferated, especially in the field of new technologies like the ICTs and biotechnologies. Many of them resorted to securing capital from venture capital finance institutions in order to start up their initiatives, some of which were not aimed at basic scientific research but above all at technological research. From the 1980s on, the size of R&D firms – some of which had already become RDI firms through the creation of corresponding departments – ceased to be important. What was important was their capacity to innovate and penetrate the market of new technologies. A few years later, this resulted in the creation of a new stock market indicator, the NASDAQ, in which technoscientific companies found a new source of financing and stock market capitalization. Most of those small RDI firms perished or were absorbed by large corporations, but some survived and went on to become major technoscientific companies. There were still government-funded macro-projects and therefore Big Science still existed. But this aside, a new form of science emerged whose research prioritized technological innovation. The size of the projects, teams, and instruments in question was not important in the case of technoscientific companies. What was decisive was their ability to innovate and for that reason Schumpeter's theories reemerged with some force during this time. This is one of the reasons why I differentiate between macro-science and technoscience properly speaking. Some small companies (Apple, Microsoft, Intel, Cisco, Google, and so on) demonstrated a much greater innovative capacity than the large postwar industrial corporations. They enjoyed extremely high growth rates although much of this was ephemeral. Innovative companies typically present a high mortality rate, but successful ones are extremely profitable.

Technoscience created financial sectors in which good, swift business could be done if one could manage to innovate. The stock market, private investors, and major financial institutions were all attracted to this new sector, leaving the macro-projects to government agencies. Overall, this new scientific-financial policy managed to reverse the percentages of public and private funding of research. The dominance of private investment thus became, from that moment on, a structural component of the American RDI system and one that other countries would try and imitate.

From an axiological perspective one could say that the most characteristic values of capitalism entered into the very heart of scientific-technological activity. Now one of the primary objectives of technoscientific companies was to make people who had traditionally been distant from the scientific community rich in a short space of time. Capitalization on the stock market and investor confidence became dominant values for many firms interested in science. Although the classic values of science were still present when it came to research, creating knowledge was not an objective for RDI companies; rather, they focused on technological innovation and its capitalization in the market. The relative importance of technical, economic, and business values grew considerably while the political values of the World War II era diminished. Moreover, many technoscientific companies became multinationals, going beyond the American market and thus becoming more aware of cultural, environmental, and social values that needed to be sufficiently addressed in order to penetrate global markets. Likewise, legal values also became more important insofar as such companies had to secure knowledge property, the management of patents, and licensed use of technological artifacts. Above all, their primary objective was now innovation—quick innovation that offered immediate returns to investors, whether it was later sustainable or not.

To my mind, a decisive factor in the success of the private technosciences during this era was a disruptive technological innovation, the ICTs. For that reason, paraphrasing Galileo, one could say that just as mathematics was the language of the modern sciences so computer science is the language of the technosciences. I will return to this point in the final section of this chapter. Likewise, the ICTs were responsible for the emergence of new markets that gradually became global in scope, beginning with new financial markets and interconnected stock markets. If, as Schumpeter used to say, opening up new markets is one of the main forms of innovation, then the ICTs contributed significantly to this form of innovation, which is not exactly technological but was facilitated by a new technological system. In sum, the development of this second phase in the Age of Technoscience would not have been possible without the ICTs, and for that reason these technologies make up another distinctive feature between science and technology. Another important novelty was the emergence of technoscientific companies, which I will comment on briefly here.

The connection between science, technology, and companies intensified greatly during this second phase of technoscience to the point that the production of scientific and technological knowledge became a new economic sector, giving rise to what is usually termed the knowledge economy. Newly fledged companies now compete in the global technology and knowledge markets, and they do so by producing, distributing, and selling a new type of commodity: intangible goods or computer-generated products based on knowledge. Such firms may be public or private, industrial or computer based, large or small, and local, regional, national, or multinational. That said, they are all guided by the innovation imperative as a primary goal and that is why they can be termed technoscientific companies, whether they are industrial or not. Simultaneously, laboratories and scientific research teams vie for public projects and contracts with companies, and all of this in order to produce knowledge, seeking niches in public or private funding for their activities. Achieving, managing and making patents resulting from profitable RDI have become a basic component of technoscientific activity and as important as research itself. Furthermore, new forms of exploiting knowledge property and making it profitable have also emerged: user licenses, franchises, access and connection subscriptions, and so on. Many of the assets that such companies have could be described in terms of the knowledge they possess or that they are able to produce, manage, and commercialize.[100] People speak about this in terms of "intellectual capital," and it is taken for granted that investment in such capital will grow and be profitable for investors. Meanwhile, it is no longer sufficient to just produce knowledge; one must now know how to present it and sell it, or put another way, commercialize it. This is required both when it comes to proposing promising research projects and when results are presented. This management and marketing of knowledge forms part of a technoscientific company's basic activities. Whether it is a case of public, private, or mixed companies, business models have been introduced regarding the organization of work and the management of technoscience, in contrast to classic academic communities that remain stuck in an outdated form of knowledge production, distribution, and management. As one can see, there has been a radical change in scientific activity, which is why it is worth distinguishing technoscientific from scientific activities, although they are closely related.

One additional consequence of this change is that technoscientific results have become a commodity and, instead of being communicated freely and publicly in specialized journals, have become private property during the initial stages of research. Greater deliberation about economic values within the axiological nucleus

---

100. For example, data, which has become an extremely valuable commodity.

of scientific activity has resulted in systemic changes in technoscience values. During the emergence phase of macro-science, there were numerous conflicts, given that "instead of exploring new phenomena, the physicists found themselves increasingly spending their time searching for ways to pursue patentable ideas for economic rather than scientific reasons."[101] Yet from the 1980s on, scientists and engineers themselves, some of whom became shareholders in the companies in which they worked, internalized these values. RDI companies are interested in scientific discoveries that are immediately publishable because this has a positive effect on the prestige of the firm (marketing), but they are much more concerned with getting patents and leasing contracts out of developing their research projects, in such a way that their R&D investment is recovered, innovations are generated, and knowledge becomes economically profitable. The coming of private investment into technoscience brought with it the imperative of knowledge profitability. In most cases, "patentability" now prevails over "publishability," reversing one of the classic values of modern science. The potential achieving of patents is one evaluation criterion in the very design of technoscientific projects, together with the ability to transfer results to companies that function in the market. Technoscience does not just evaluate epistemic impact (publications, citations, and so on) but also and above all the economic effect of the resulting innovations, as well as the general capacity to obtain funding to develop projects. Technoscientific culture thus contains a strong business component, which was not the case with modern science, with a few exceptions.

Once more, then, we see that in the transformation from science to technoscience there was a profound change in values. In large part, technological communities had internalized business principles and values during the industrial age. During the era of macro-science, scientists collaborated on large-scale military projects for epistemic reasons (to resolve scientific problems) but also for political motives (patriotism, the defense of democracy, and so on). Now, however, many scientists have embraced business values without relinquishing their specific episteme values. Here, too,

101. Galison, "Introduction: The Many Faces of Big Science," 4.

in this case, there are differences of degree, given that some technoscientific firms tend to turn into great holding companies that are quoted on the stock market or become part of major finance groups. Marketing has become a typical practice of technoscience. While it is mostly designed and shaped by marketing experts, scientists and engineers have also emerged who stand out for their ability to "sell" or disseminate a product. This process is mostly evident in the private sector but can also be seen in publicly funded science. The increasingly close links between universities and business are a good indicator of this.

Technoscientific Activity and Values

From an axiological perspective, the second phase of technoscience has involved a profound transformation. Economic, business, and capitalist values have increasingly permeated technoscientific activity and been integrated into the axiological nucleus of technoscience research, teaching, and application. It is important to underscore this fact because from this point one infers that the axiology of technoscience must bear in mind at least three value systems: epistemic, technical, and economic. The terminology typically used to discuss these is research, development, and innovation, with the latter alluding to the business components of technoscientific activity. The new dimension of all this is that while epistemic and intellectual values (knowledge, basic scientific research, and so on) prevailed in Bush's linear model, knowledge is subordinate to innovation in the second phase of technoscience. The classic values of science have ceased to be mean values — a goal to be converted into values — in the sense advanced by Max Weber.

Put bluntly, although there may be subsidiary values, *technoscience is always guided by economic values*; something that happened only occasionally in the case of science. Economic criteria are one of three axiological components that guide technoscientific actions and are applied *ex ante* and *ex post*. For that reason, axiological pluralism is "innate" to technoscience. Some classic sciences could have been guided by exclusively or predominantly epistemic values. Yet this is not the case with technoscience, and we therefore have a new axiological criterion to distinguish the sciences from the technosciences: the existence of an economic values subsystem alongside the previously mentioned subsystems of epistemic and technical values. I will return to these axiological considerations in chapter 4, but at this stage it is worth introducing some of the principal ideas involved therein as a means of demonstrating the systemic and pluralist condition of the axiology of technoscience.

First, epistemic values continue to be significant for technoscience because their innovations and proposals are based on previously corroborated scientific knowledge. Current technological artifacts are usually constructed on a foundation of sufficiently corroborated scientific theories and contributions, which gives greater reliability to the subsequent functioning of such apparatuses, because they follow the laws of nature and because different situations have been analyzed. A good example of this can be found in pharmaceutical products, which include explanations about their contraindications and possible harmful secondary effects. Therefore the internal values of science (objectivity, intersubjectivity, credibility, prediction, verifiability, refutability, empirical suitability, precision, rigor, coherence, productiveness, generality, observational repeatability, measurements, experiments, and so on) are reflected in the technological artifacts themselves and not just in the theories used. Several of these values form part of the axiological nucleus of technoscience.

Second, the typical values of technology are among those values underlying technoscientific activity: novelty, functionality, efficiency, effectiveness, utility, applicability, reliability, feasibility, simplicity of use, speed of functioning, durability, versatility, integrability with other systems, and so on. These criteria are extremely influential when it comes to evaluating technoscientific proposals and activities. Note how many stem from properties possessed by technological systems; this transformation of properties into values is a frequent phenomenon in the axiology of techniques field. Plurality also reigns in this second form of values, even though efficiency, utility, and effectiveness may be considered priority technological values.

Third, certain economic values became very influential during the second half of the twentieth century. Such values include the intellectual property of knowledge (patents), productivity, competitiveness, the maximization of gains, profitability, the optimization of resources, management quality, cost reduction, marketability, and transportability; most of these are important for entrepreneurs who promote innovation processes from their R&D departments. Much of current scientific research is financed by business, so it is hardly surprising that economic and business values increasingly permeate technoscientific activity. The science economy forms part of STS studies and makes interesting contributions when it comes to defining the notion of technoscientific rationality.

Fourth, given that technoscientific activity is developed in social frameworks that are legally regulated, a subsystem of legal values emerges that companies must respect and comply with: property, legality, security, publicity, adaptation to current regulations, and so on. This is because there are frequent legal conflicts that are settled in court rather than in the scientific communities themselves. Competition among technoscientific companies is not just resolved in markets but also before judges. And often the courts decide who innovated and who did not. Nowadays, any technoscientific company must have a legal department to defend its interests, for example in disputes over patents or user licenses. There was conflict in modern science over issues such as who discovered something first or plagiarism, but it was the scientific communities themselves that settled such disputes. Current technoscience gives rise to much more complex conflicts that are resolved by authorities outside science: parliamentary commissions, juries, ethics commissions, and so on. Evaluation of the technosciences, and in particular of its risks, is undertaken on the basis of hybrid value systems.

Fifth, the impact of industry and new technologies on nature has led to intense reflection on the risks associated with certain technological innovations, beginning with weapons and nuclear energy, with the concomitant emergence of new values that could be generically termed environmental. These would include, for example, non-contamination, conservation of the environment, biodiversity, risk evaluation, sustainability, and the minimization of animal suffering. Some radical defenders of environmentalism even speak in terms of ontological values in the biosphere.

Sixth, the effect of new technologies on everyday life and society has brought a series of social values (intimacy, privacy, solidarity, liberty, stability, multiculturalism, cooperation, literacy, and so on) to the fore, and these allow us to value the social impact of technoscientific activity. This sixth group is closely linked to the fifth and seventh groups, respectively, with the result that they often have to be analyzed together.

Seventh, there are political values that are also significant in technoscientific activity and specifically in its planning by means of science and technology policies. These include research freedom, democratic values, equality of opportunities, control of public spending, power relations, tolerance, the right to disagree, and especially, prudence, which often leads to moratoriums being established on some areas of research.

Eighth, the first phase of technoscience (Big Science) was closely linked to military activity and it continues to be so today. Typical technoscientific projects such as the ENIAC, the Manhattan Project, space exploration, and more recently, nanotechnologies, were sponsored by the science-military-business lobby that shaped the development of technoscience in the United States. Such initiatives take into account military values such as patriotism, victory, the right to defense, secrets, hierarchy, discipline, obedience, and so on. The two world wars and the Vietnam War induced profound moral crises within some scientific communities that should be interpreted as conflicts provoked by the sudden appearance of a new subsystem of values within the axiological nucleus of technoscientific activity. This eighth subsystem of values is still important, moreover, in times of peace.

Ninth, esthetic values (beauty, elegance, simplicity, harmony, and so forth) also affect scientific activity insofar as representations of knowledge emerge that must be attractive and, therefore, allow for esthetic evaluations. In the formal sciences (mathematics, logic, computer science, and so on) such values are particularly important although they are also present in other sciences: one only need cite the example of Einstein and his theory of relativity, in which the symmetry of equations is valued very highly as a principle of ordering the world. The current techniques of scientific visualization by means of computer simulation in two or three dimensions are a good example of the hybridization among the arts, sciences, and technologies.

Tenth, some technological developments give rise to profound ethical and religious problems in such a way that current technoscience is also influenced by moral and religious values, especially in the case of the biotechnologies. In recent decades bioethics, and more generally science ethics, commissions and studies have emerged everywhere, something that did not happen during the modern science era. I will not keep on about this point here, but suffice it to say it is another of those arguments that reveals the profound change experienced by science during the final third of the twentieth century. For my part, I tend to believe that ethical values are secondary in technoscience: they are not to be found at its axiological nucleus although they may be significant in specific questions. Such is the case with the existence of ethics committees in different technoscientific institutions and companies. Obviously, values such as life, health, human dignity, duty, autonomy, responsibility, veracity, honesty, altruism, freedom of conscience, respect for beliefs, tolerance, and many more are important in different contexts of technoscientific activity. The honesty of scientists, for example, is a condition sine qua non of technoscientific activity and therefore a central value. However, moral and religious values are not the only values in technoscience and occasionally are not even the most important ones when it comes to decision-making. For that reason, the axiology of technoscience cannot be reduced to an ethics of science.

This initial typology of ten axiological subsystems can be improved on and refined, but it does at least allow one to illustrate the amplitude of the axiology of technoscience. Philosophers of science tend only to be interested in epistemic values, moral philosophers in ethical questions, soldiers in victory and the means to achieve it, economists in the cost-benefit relationship, jurists in respect for the law, and environmentalists in defense of the natural world. All these analytical perspectives are valid, but none of them eradicates the axiological problems generated by current technoscience. Indeed, there are just as many other axiological dimensions of technoscientific activity. Precisely for this reason, it is necessary to address the problem of axiology in as general and widespread a way as possible, instead of reducing the issue to one of the ten analytical frames mentioned above. Axiological pluralism is one of the structural features of technoscience and, for that very reason, a solid starting point for the philosophy of technoscientific practice. Any in-depth study of technoscience must bear in mind many other factors and facets, but axiological analysis allows one to examine quite deeply the structure of contemporary technoscientific practice.

The Social Technosciences

Up to now, I have referred mostly to the physical and natural technosciences, plus a brief incursion into technomathematics in my discussion of the ENIAC. To conclude this chapter I will extend that reflection on technomathematics and introduce the issue of social technosciences. My goal here is to differentiate these from the social sciences, a topic I will revisit in the following chapter.

Mathematical models have a very important function in modern science: on the one hand, representing phenomena and data, and on the other, expressing scientific laws. Mathematics has contributed formal languages that were later interpreted in different semantic ways. Interpreting a formal language consists of assigning meanings to its symbols in such a way that the formulas that express axioms, laws, and properties are true. In this way, mathematical formulas became the canonical expression of scientific laws, thanks to the universality, precision, and rigor of their formulations. Mathematical laws that express the scientific laws of a theory – for example, Newton's laws – are not limited to describing the trajectory of important things. Moreover, these trajectories verify the corresponding equations, which contributes real value to empirical principles. Scientific laws explain and predict the movement of important things precisely because mathematical formulas produce true principles in that they are interpreted in empirical terms. This is the power of verification and explanation as opposed to simple descriptions and statements. That said, it is not easy to achieve a mathematical model that is verifiable in all its details within an empirical domain; it is more typical for models to offer approximations of the empirical data.

The spread of computers and ICT technologies in the second half of the twentieth century facilitated the emergence of a "new kind of science," to put it in the terms of Stephen Wolfram, designer of the Mathematica software program. According to him, "three centuries ago science was transformed by the dramatic new idea that rules based on mathematical equations could be used to describe the natural world; my purpose in this book is to initiate another such transformation, and to introduce a new kind of science that is based on the much more general types of rules that can be embodied in simple computer programs."[102] Computer simulations have superseded analytical mathematical models. When reasonably complex systems are studied, their behaviors cannot be reduced to mathematical formulas, or even to systems of equations and formulas, precisely because they are phenomena with degrees of non-polynomial and even exponential complexity. Instead of formulas, one must use different algorithms to represent and interpret such systems, for example in the dynamics of complex systems. Bush, Zuse, von Neumann, and many others designed their computing machines in order to address physical problems (missile trajectories, explosions, critical masses, and so on) that could not be analyzed by algebraic or differential equations, and some occasions not even by statistical models, owing to the huge amount of data that had to be handled and the difficulty of computing this through traditional means. One of the first successes in the emerging technomathematics was the Manhattan Project, in which the Monte Carlo algorithm was used to calculate the interactions among twelve hard spheres that were impenetrable – a question that could not be approached through analytical methods. Since then, computer methods have been applied to multiple issues (cosmology, elementary particles, fluid dynamics, wind tunnels, flight simulation, population dynamics, ecosystems, meteorological forecasting, climate change, macroeconomics, market studies, cognitive sciences, sociological surveys, migratory

102. Wolfram, *A New Kind of Science*, 1.

flows, and so on), giving rise to the distinction between "analytical methods" and "computational methods." This is similar to the distinction I now propose between mathematical models and technomodels. Numeric (or technonumeric in my terminology) models represent phenomena and data on computer screens, but they do not carry out the second aforementioned function, namely formulating scientific laws, because they have different objectives. Technomodels are not constructed to explain phenomena but instead to represent possible situations by modifying the corresponding parameters. Their function is above all heuristic although it can also be predicative, but without relying on the deductive-nomological model. Changes of state in physical, biological, social, and economic systems are represented by bit sequences, which allows one to control subsequent changes and simulate what would happen in a given initial measurement condition. Technomodels do not explain what might happen because they function in uncertain conditions, but they are capable of representing what can happen in given circumstances. This *simulation of the possible* is more than sufficient in the case of the social sciences because this discipline provides a sufficiently empirical foundation to make decisions based on data, even if such data are simulations of facts rather than scientific facts in the traditional sense of the term.

The different epistemic function of technomodels is due to the fact that the ultimate goal of the technosciences is not to acquire knowledge with a true value but rather to efficiently transform the data, phenomena, and systems studied with the aim of being able to operate with and modify them, all while maintaining a certain foresight of what can happen if a given action occurs. As regards the social technosciences, they deal with very complex systems and operate in non-determinist and even uncertain conditions. Computer simulations, both because of the transformative tendency of technoscience and because of the conditions of uncertainty of the objects and processes under study, have been a very powerful tool in the study of such phenomena. The ICTs have demonstrated their great capacity to modify variables and parameters, as well as algorithms, adapting technomodels to the most diverse situations. Thanks to computers and computer simulations, the technosciences have been able to address very complex scientific, technological, economic, and social problems that were not within reach of traditional mathematical models. This would have been impossible without the ICTs, and for that reason it is appropriate to term this new mathematic model a technomodel, due to the fact that technologies are essential to its operation. Scientific research has always relied on scientific instruments, but in the case of the technosciences there are no simple instruments but, rather, true mediations of scientific knowledge.

Scientists still use mathematical models, especially in order to represent scientific theories and laws, but besides this they also increasingly use technoscientific models facilitated by the ICTs, of which computer simulations are an important example. Two significant examples of technomodels are: on the one hand, databases, code systems, and numerous cryptographic technologies, and on the other, interconnection protocols among computers that allow the Internet to function. All these examples, and others that could be cited, are possible thanks to the ICTs, but the concept of simulation is not the most appropriate one to study them. I prefer the term "technomathematics" (and "technomodel") to "computing simulation," although one can still use the latter without any problem at all given that it is a subclass of the technomodels. Given that scientific practice was transformed radically in the final decades of the twentieth century, the philosophy of technoscience must introduce new conceptual analysis tools such as, for example, the notions of "technoscience" and "technomodels." Many of the models produced by the ICTs cannot be considered mathematical models because they are not interpretations of formal languages. Traditional categories of the philosophy of science are inadequate for technoscience.

It is not easy to define the notion of a model, as various observers have noted,[103] owing to the great plurality of models that scientists use. However, Stephan Hartmann offers an interesting definition of computer simulations: "I maintain that the most significant feature of a simulation is that it allows scientists to *imitate one process by another process*. 'Process' here refers solely to a

103. See, for example, Agassi, "Why There Is No Theory of Models."

temporal sequence of states of a system."[104] For my part, I would prefer to speak about representation rather than imitation, but I will not enter into this debate. For the time being, I will accept Hartmann's definition. In line with his definition, any physical, biological, economic, social, or cognitive process, for example, can be simulated by computers, given that digitalization and programming languages make the numeric simulations of a specific object or state change, in turn, state, which is manifested perceptively as a modification of the images that appear on the screen. Therefore, technomodels are not mathematical formulas or formal languages but rather bit sequences that are represented by artificial images and sounds that resemble the objects and processes under study, whether in physics, chemistry, or the social sciences. Once constructed, technomodels are controllable and operable in such a way that we can represent natural and social processes by means of their corresponding digital processes, which in turn change according to the passage of time. This allows for representation of the most diverse situations possible, not just those that in effect occur. Without being predicative in the deductive-nomological sense, technomodels represent the evolution of different types of systems, as well as their changes of states, independent of the complexity of the phenomena under study.

Note that the idea of technomodels is the same for both the natural and the social sciences, which is important for the philosophy of science. In fact, there are also technomodels that are used in mathematical research, for example to calculate zeroes in the Riemann zeta function and verify empirically its conjecture, which is only possible thanks to computers. Note, too, that there are many varieties of computer languages, as many as there are configuration, programming, and processing systems. Digital writing is technowriting, and for that reason the models it generates must be termed technomodels.

The interesting feature of technomodels stems from the possibility they afford to fully control computer (computational) processes that take place inside computers. Modifying parameters and algorithms, one never knows what is real. Nor can one ever formu-

104. Hartmann, "The World as Process," 77.

late scientific laws, but one can foresee to some extent what would happen in digitally represented systems in a given circumstance. Not much else is necessary in order to make decisions in the social sciences and, therefore, to prefer some representations as opposed to others. The different alternatives can be evaluated, yet not according to the values of truth of the model theory but rather according to other values such as precision, the effectiveness of algorithms, complexity, productiveness, empirical suitability, and so on. In sum, technomodels and the technosciences refer to other kinds of values that are not just epistemic but also technological, economic, political, and social.

Technomodels can also refer to scientific activity, which can be conceived as a system of actions and courses of actions.[105] This is why technomodels are valid in order to analyze and reconstruct scientific practice and not just scientific knowledge. A technoscientific agenda can be conceived as a family of sequences of actions, each of which can be represented by the appropriate technomodels. The philosophy of technoscience revolves around the idea of a technoscientific agenda, and its structure can be analyzed by different technomodels, each of which simulates some of the basic technoscientific actions.

The same is possible in the case of innovation processes. These are not predictable because they are nearly always produced in conditions of uncertainty. The social sciences find it difficult to explain innovation processes. However, the recent emergence of the social technosciences offers a new method of researching innovation processes, based on computer simulation drawn up from empirical data. My next task will be to examine these different systems of indicators that institutions such as the OECD, Eurostat, and Innometrics have created in recent decades in order to analyze R&D systems (the *Frascati Manual*) and, more recently still, innovation systems and processes. RDI studies must use computer simulations and technomodels to analyze and reconstruct technoscientific practice and in particular innovative activities, which constitute the ultimate objective of technoscience.

105. See Echeverría,2007b.

2
THE DOMINANT PARADIGN IN INNOVATION STUDIES: THE OSLO MANUAL

The Emergence of Innovation Policies

Following the sudden change implied by the appearance of Big Science in the United States, there was a second structural transformation in the 1980s in accordance with the evolution of R&D to RDI systems. Innovation thus became the main purpose of R&D systems, which gave rise to considerable changes in the structures of these systems as well as in science and technology policies. In order to be up-to-date, technoscientific teams must now generate not just knowledge, reflected in high-impact publications, but also innovations. Given that companies innovate, research groups must forge links with such firms. Research in itself is not enough.

This change can be expressed by saying that scientific communities have ceased to be the agent subject of science, having been replaced in that role by technoscientific companies whose strategies are guided by the imperative to innovate. Handling scientific research in a businesslike way and arguing that innovation is the ultimate goal of research are two distinct features of current technoscience.

Meanwhile, the concept of innovation has been extended in recent years. The original technological innovation heralded by the neo-Schumpeterian school has been joined by service, process, marketing, organizational, user, educational, urban, and most recently of all, social and open innovation, of which the latter two will receive special attention in this book. With all of this, of course, one cannot overlook cultural and artistic innovation (creativity) as well as political, juridical, and military innovation, could also be discussed at length; here I will limit myself to just a brief mention of them. As RDI systems have been developed, so the concept of innovation itself has become more complex and difficult to explain. Castells argues, "Knowledge and innovation are sources of wealth, power, and quality of life."[1] And these words capture the profound change experienced by industrialized economies and societies insofar as they have created information and knowledge societies.

In order to defend his thesis, Castells offers a particularly incisive example: "from knowledge comes money; with knowledge one ends up achieving investment and earning money; without knowledge, although one may have money, one ends up losing."[2] This is true, but with two provisions: First, knowledge and innovation do not just serve to generate wealth, power, and well-being; they also destroy wealth, overthrow previous forms of power, and significantly diminish the level of well-being of wide sections of society, as Schumpeter rightly contended. The financial crisis that broke out in 2008 and produced a true debacle in the international stock markets originated in a few "financial innovations" that created subprime mortgages and other novel financial products, but which shortly thereafter proved lethal for many banking institutions.

Second, knowledge cannot be reduced to scientific knowledge alone. There are other forms of knowledge that are just as important, and sometimes even more important, when it comes to generating wealth, power, and well-being. Being in possession of privileged information, for example, is typically very important in the world of business. Such information hardly ever originates in experimental laboratories (although sometimes it does, when discoveries are made for instance) but usually implies other forms of knowledge (the concepts of know when, know who, know where, and so on). Such other forms of knowledge are rarely discussed

---

1. Castells et al., *La transición a la sociedad red*, 20.
2. Ibid.

because there is a tendency to identify knowledge with science and technology. The Schumpeterian entrepreneur senses opportunity (the Ancient Greek notion of *kairos*) and knows by institution who he can trust and who he cannot. Many innovation processes have been developed on a foundation of trust (know who) beyond R&D systems. One must, therefore, distinguish several different types of knowledge and analyze empirically which of these is decisive in each innovation process, dispensing with the dogma that innovation only results from R&D. I will demonstrate in this work that there is a range of different kinds of knowledge and forms of innovation, contrary to the argument implied by the linear model that, besides being linear, is both monist and reductionist: namely, that all economically important knowledge ultimately stems from R&D. Such an argument is incorrect, as I will show in chapter 4. I agree that knowledge is the source of wealth, power, and well-being, but the issue is: what kind of knowledge? When people speak about a knowledge society, one usually just thinks in terms of scientific and technological knowledge. This is another dogma I take issue with in this book.

In this chapter, I will analyze the dominant paradigm of innovation studies whose principal expression is the OECD's *Oslo Manual* and Eurostat. I will show that, as the concept of innovation has been expanded, so this paradigm has been shown to be insufficient, including its system of indicators. Recently, several forms of hidden innovation have been identified that are not recognized or measured by the indicator systems of the *Oslo Manual*.[3] A crisis has thus beset the dominant technoparadigm in that many innovation processes remain outside its field of analysis. This has resulted in the emergence of an alternative paradigm, social innovation, which I will address in chapter 3.

Evolutionary Economics and Innovation

One of the most influential philosophical tendencies in economic studies of innovation finds inspiration in biological evolution. It is based on the "evolutionary economic theory" devised in the

3. For more on this, see chapter 4.

early 1980s by Richard R. Nelson and Sidney G. Winter,[4] which is derived clearly on Lamarckian principles. Evolutionary theory in economics conceives of markets as economic systems that select the most suitable companies, those that best adapt to the environmental circumstances in which they carry out their business. Technological innovation is a key part of this, and there were initially two variants: product innovation and process innovation. Product innovation gives a company an advantage over its competitors because it involves launching new and until then inexistent merchandise on the market. Process innovation implies altering a company's production processes so that productivity, and therefore competitiveness, is increased. As a whole, the more a company has the capacity to innovate, whether with products or processes, the more likely it is to survive.

This evolutionary notion of innovation dominated economics through the end of the twentieth century. Indeed, it was a conceptual cornerstone when it came to drawing up the first *Oslo Manual* in 1992. Schumpeter was a big influence on proponents of this idea, to such an extent in fact that Nelson and Winter are considered neo-Schumpeterian. One should mention here that Nelson and Winter do not apply a Darwinian model in their theory, in contrast to other theoreticians of technological change who try and extend the theory of evolution in order to explain processes of technological change. Nelson and Winter consider markets and economic systems to be "selection environments," but the struggle for survival in these environments is not governed by chance, although it does always take place in conditions of uncertainty. Jon Elster addresses this issue and points out four differences between Nelson and Winter's evolutionary economics and a possible Darwinian theory of innovation:[5]

1. According to Nelson and Winter's idea, there are no equilibria, except in a few specific cases, in an economic system, coinciding on this point with Schumpeter's argument.
2. Variations are not random but instead the result of intentional search processes conducted by economic agents, en-

4. Nelson and Winter, *An Evolutionary Theory of Economic Change.*
5. See Elster, *Explaining Technical Change*, 138–50.

trepreneurs.

3. Companies do not reproduce themselves but instead increase or decrease in size.
4. The selection environment can change very quickly, which means that at any given time both efficient and inefficient companies exist side by side.

I will return to these issues in chapter 5 when I propose a naturalized conception of innovation, yet not one that is Darwinian but, rather, pluralist. As regards the issue of change and technological innovation there is much to say, given that not all change implies innovation. To my mind, in order to explore this matter in more depth one should turn to an axiological analysis. The *Oslo Manual* makes it clear from the outset that innovation favors the growth and productivity of a company, thereby making it more competitive in markets: "it is widely accepted that innovation is central to the growth of output and productivity."[6] Companies try to grow and survive in markets, and in order to do so they must compete with each other, on the assumption that we are analyzing innovation processes in a market economy. Innovation is subordinate to a system of economic values, some of which are very significant such as the growth, productivity, competitiveness, and survival of the company itself. Business innovation is not a goal in itself but rather a means to achieve entrepreneurial objectives that are considered beneficial and that even make up the axiological nucleus of capitalism and the economies of the free market. In principle, the same is true of biological evolution: it is just a means of favoring the survival of a particular species. According to Jean-Baptiste Lamarck, the important thing is that the species adapts to the medium, or Nelson and Winter's selection environment. For any company, the canonical expression of this selection environment is a market, but also public R&D policies, which firms and technoscientific agencies must also adapt to.

Nelson and Winter, proponents of evolutionary economic theory, are considered neo-Schumpeterian in innovation studies, and they use Lamarckian-inspired models in their theory of economic evolution. In their opinion, not all technological change implies

6. OECD and Eurostat, *Oslo Manual*, 10.

progress or innovation; there is technological innovation when a new process emerges that technically improves a previous process, or when new processes with other inputs replace previously existing processes in a more efficient way.

Furthermore, these authors also accept the Schumpeterian distinction between invention, innovation, and diffusion. Inventions stem from basic and applied research, but one should not confuse them with innovation. An invention becomes an innovation when a company transforms knowledge into an economic commodity, whether as a result of making new products that are successful in the market or as the consequence of introducing a more efficient method of production through reduced running costs or greater productivity. For this to take place, there must be profound knowledge of the selection environment, a condition that allows one to distinguish between inventors and innovators. The former, who may be creative and even visionary people, are not typically overly concerned with the commercial viability of their inventions. The latter, however, know very well how to discern what is possible in a given economic setting; moreover, they know how to place value on the invention. This is a quality associated specifically with innovative people and agents.

According to evolutionary economists the selection environment is always made up of competitors, consumers, providers, the institutions devoted to creating human capital, and regulatory institutions like states and governments. Extensive knowledge of the market allows one to understand the services that consumers demand, to define what provisions a new product must provide, and to have a good understanding of existing legislation and R&D policy priority areas in order to adapt any innovation to these conditions. Technoscience has something to do with markets and industries, but also with public policies, laws, and users of potential innovations. For me, technoscience is characterized by the fact that it is developed in mixed selection environments in which different evaluation criteria operate, not just the results of competition in markets. This means that it is important to examine the different evaluation criteria that area used to favor one innovation over another: they are not purely scientific, or purely technological, or purely entrepreneurial, or purely legal, or purely political. This is why I am going to propose a pluralist model of hybridization, in which mixed or hybrid systems of evaluation operate.

While Nelson and Winter's systemic model represents an important advance in comparison to the linear model, throughout this work I will attempt to demonstrate that section environments have a much more complex and varied structure than those they themselves employ, both as regards the sources of innovation and in terms of their types and scales. Nevertheless, I fully accept one of the assumptions of these neo-Schumpeterian evolutionists: without acceptance or diffusion, there is no innovation or successful invention. Many inventions throughout history have been halted in their tracks, stuck in the world of the possible, without ever taking shape in the real world. In contrast, innovations must take that step from the possible to the real, while at the same time competing with other innovations. In order to be diffused (to grow and extend) they must demonstrate that they have improved on what came before, or at least that they are preferable for whatever reason. There are no successful innovations without some form of positive evaluation of their new contribution, of whatever kind it may be. By means of an advance statement of the basic argument of this book: there is no innovation without a prior evaluation process. According to neo-Schumpeterian economists, markets evaluate different innovative proposals, selecting some and ruling others out. For my part, I would argue that there are other scenarios, not just markets, in which innovative proposals are also evaluated. Innovation is related directly to values, and generates value — but not just economic value. And clarifying the relationship between innovation processes and values is one of the aims of this book.

Turning once again to neo-Schumpeterian evolutionary economists, they are interested solely in *companies that adapt well to and are successful in the economic environment on the basis of their innovative proposals*. They thus presuppose a Lamarckian adaptation process that may or may not be competitive and that does not require a static equilibrium. Adaptability is a priority value but does not necessarily imply a competitive situation. Nor does it require absolute adaptability to survive, with a certain degree of adaptation being sufficient. This would be the definition of an adapted company that best captures the evolutionary approach. Innovative companies are those that best adapt to the market of a given situation, according to evolutionary economics, but that does not imply maximizing any particular value or even competitiveness. For my part, I will try to show adaptability is necessary but not sufficient in itself. In other words, a Lamarckian theory of innovation is insufficient, as I will suggest in chapter 5.

The Social Technosciences of Innovation

According to evolutionary economists only companies capable of innovating survive. Innovation is a necessary condition for survival in the market. One must understand the innovative capacity of every company, and therefore one must measure innovation, both in terms of its immediate and medium-term results. Business policies must be geared toward empirical data that allow one to compare a company with its competitors. The same is true of regions and countries. In accordance with these postulates, the OECD took on the task of measuring innovative capacities and comparing the results, later extending the analytical perspective to include not just companies but also regions and countries. However, it is not the same to measure the economic benefits of a company or its share of the market and its innovative capacity. Measuring innovation is a difficult and complicated task that has required the creation of specific instruments for the undertaking.

Before discussing the *Frascati Manual* and the *Oslo Manual*, it is worth thinking for a moment about the methodological and epistemological challenge that measuring innovation processes implies. One thing is to measure R&D results, which are normally expressed in the form of scholarly publications and patents; another is to weigh up the results of an innovation process, which are never obvious or transparent due to the fact that many different agents with conflicting interests intervene in such processes. For the same reasons, it is not always easy to specify exactly who the innovative agent is. According to Schumpeter, the innovator is a person and specifically an entrepreneur, although he also admits there is no innovation without a subsequent process of social diffusion that depends on many variables. Measuring the results of an innovation process is a complex operation because of the fact that, following the immediate results, unpredictable consequences may be derived that call into question the initial results and offer a quite different analytical perspective. Put simply, anything new can be beneficial for some and harmful for others, and for that reason innovation is creative destruction. Now, if the results are not certain, it is difficult to undertake an "objective" analysis of an innovation process. The most one can do is compare some innovations to others, establishing which have been more and less successful in a certain period of time. In fact, the *Oslo Manual* does not even measure economic output, limiting itself to checking whether a firm is promoting innovation activities or not at a given time. Innovations cannot be arranged in linear order, nor do they possess any dimensional unity that would allow one to determine the greater or lesser success of one as opposed to another. Ultimately, this is not an issue that can be approached from the social sciences, but rather from another form of knowledge: the social technosciences.

Science attempts to understand what the world and, where applicable, society is like. Technoscience, however, attempts to transform the world, including social worlds. The data obtained in innovation studies are proactive, given that they shape actions, strategies, and policies to be undertaken. These are not empirical facts but data that shape subsequent actions whose evolution and results cannot be predicted. In some cases, performative concepts and methods are used that define what innovation is according to the procedures measuring it. As the subtitle of *Oslo Manual* indicates, these are "guidelines for collecting and interpreting innovation data." In other words, it specifies how to obtain data on innovation and compare the data obtained. Yet innovation is extremely complex and ultimately we do not really know what it is;[7] the most we can hope to do is measure it, that is, compare some innovative processes and agents with others. Innovation studies within the dominant paradigm focus on *knowhow*, on how to collect and

7. As stated quite clearly at the beginning of the *Oslo Manual*: "while our understanding of innovation activities and their economic impact has greatly increased since the first edition of the Manual, it is still deficient." OECD and Eurostat, *Oslo Manual*, 10.

interpret data, instead of confronting the main conceptual challenge of examining what innovation is. This is one of the reasons why the concept of innovation as studied by the *Oslo Manual* is a technoscientific more than scientific concept. Let us examine this briefly.

Philosophers of science distinguish three classes of scientific concepts: classificatory, comparative, and metrical. The first of these introduces criteria of distinction within a set of entities or events, in such a way that the set studied is divided into subsets or classes, which in turn may be differentiated once again on the basis of classification subcriteria. The second indicates when two objects or events that share a particular feature or property coincide in regard, or if one precedes the other in relation to that feature or property. Metrical concepts "are features of the most advanced stages in science"[8] and assign real numbers to the objects or events studied, which allows one to compare and contrast them with great precision and normally in relation to a unit of measurement. The great advantage of measuring any kind of domain is that, insofar as it is achieved, it is then feasible to carry out mathematical or statistical studies with all the possibilities that this offers. Different social sciences (economics, sociology, psychology, and so on) have managed to measure their fields of study, at least in part. And this has imparted them with somewhat of a scientific nature, more than those disciplines that just use classificatory and comparative concepts. As José Antonio Díez and Ulises Moulines point out, "every metric concept matches a comparative concept and every comparative concept a classificatory one."[9] Nevertheless, innovation is a classificatory and also a comparative concept, but one cannot strictly say it is metrical, although there is a lot of literature and several indicator systems to measure innovation. Put another way: If we measure something we can compare objects, and so long as we can compare we will be able to classify. But the reverse is not true. One can compare innovation processes, but that does not imply that the indicator systems we use to do so involve metrical concepts. For that reason, one must distinguish

8. Mosterín, *Conceptos y teorías en la ciencia*, 25.
9. Díez and Moulines, *Fundamentos de filosofía de la ciencia*, 100.

different *levels of a scientific nature* in the concepts we use, according to whether these concepts are solely classificatory, comparative, or, in the final analysis, metrical.

This is one of the first methodological problems to deal with in relation to the concept of innovation from the perspective of the social sciences. The *social sciences of innovation*, on the premise they exist at all, should use scientific concepts (metrical, comparative, or simply classificatory) to study innovation processes. Most philosophers do not try to be scientists and do not, therefore, resort to scales or units of measurement. However, economists, sociologists, and psychologists, to mention three social sciences, try to use metrical concepts in order to study innovation. If they were to manage this, it would lead to *scientific studies of innovation*, which some economists and political scientists of innovation claim to have arrived at. Others approach innovation issues from non-scientific (artistic, literary, and so on) perspectives. In such cases, the debate typically revolves around creativity, as we will see in the next chapter when I discuss literary, artistic, and cultural innovation, some of which is socially and economically significant. Moreover, there are people who are not interested in innovation studies but are interested in innovation itself because they practice it; in other words, they are practitioners, without being bothered about studying it.

This book concerns the field of both scientific and non-scientific innovation studies. I have alluded to innovation in biology, and now I am offering a few reflections on the social sciences of innovation. My hypothesis is that innovation studies, as they are at present, do not form part of the social sciences but are instead part of the *social technosciences*, with the latter understood as knowledge and techniques that aim to modify and transform social objects, actions, relations, and, where relevant, societies as a whole or at least some social groups or sectors. Put another way, the sciences examine what societies, relations, systems, and social behavior are like; the technosciences are proactive and even regulating in that they are aimed at transforming human and social actions and relations. The social sciences try to measure their objects of study and share them on the basis of units of measurement: for example, money or individuals, establishing people as a unit of measurement (one person, one vote). The social technosciences, meanwhile, also use quantitative methods but without presupposing the existence of units of measurement and sometimes not even metric scales. That is why we speak in terms of technosciences more than social sciences.

Measuring R&D: The *Frascati Manual*

Throughout the twentieth century and especially after World War II the sciences and the technosciences came to acquire a growing economic and social importance. Following the report by Vannevar Bush, *Science, the Endless Frontier*, the United States developed specific policies for science and technology. This initiative was then copied by other countries, and the R&D sector began to take shape as a new and increasingly significant economic field. Simultaneously, there emerged a need to have empirical data on the growth and evolution of this sector in different countries and, with that, studies on science and technology. The history and philosophy of science became more professional in the mid-twentieth century following the establishment of the International Union of History of Science (1947) and the International Union of Philosophy of Science (1949). The sociology of science, fostered by Robert Merton and his followers, also emerged during the same era. Given the huge investment that Big Science required and the important industrial, military, and political consequences that macro-scientific programs faced, planning such activities became a necessity. The study of the structure of national science and technology systems became an objective of scientometrics, a new scientific discipline that has grown in importance since 1950. Certain fields of scientific activity began to become measurable – for example impact factor indicators for every article and journal (the Institute of Scientific Information, then ISI-Thomson, and nowadays Thomson-Reuters) – as did economic data on investments in, the existing resources of, and the maintenance costs of R&D activities in every country. The OECD specialized in this latter task, creating an international group of experts in science and technology indicators in 1963. The result of the group's work was publication of the *Frascati Manual*, which began a series of manuals that have become international standards when it comes to measuring RDI systems and activities. These include the *Oslo Manual* (1992), which measures innovation activities, and the *Canberra Manual* (1995), which studies the number of science and technology personnel in every country.

In sum, a new social technoscience focusing on the study of science and technology (R&D) systems emerged during the second half of the twentieth century, and in the 1990s, this was expanded to incorporate the study of national, regional, and local innovation systems. This "science of science," as Solla Price terms it[10], led to certain empirical studies of science and technology and, some years later, innovation studies, based on the use of empirical indicator systems. The *Frascati Manual* and the *Oslo Manual* embody this technoparadigm in RDI studies, based on the economic and quantitative analysis of technoscientific activity. For that reason, I will pay especially close attention to them in this section.

The *Frascati Manual* is concerned exclusively with R&D, in which it includes basic research, applied research, and experimental development.[11] It advises studying it by means of internationally agreed on studies and surveys, with the aim of being able to compare data from different countries. The *Oslo Manual*, meanwhile, does much the same as regards innovation. These two requirements are typically technoscientific, given that they regulate data collection and introduce a value that would be decisive in the social technosciences that address RDI: international comparability among data. Put another way, the OECD gave instructions to social scientists about how to measure R&D, and then innovation, with the aim that such data might be useful for the science policies of the different countries that make up that organization. The knowledge sought, then, was from the outset an instrument for other ends, in this case those regarding each country's science policy. To show with empirical data that one country or region is ahead of another in R&D or innovation indicators is a political objective in its own right, especially if those data are offered by an

10. Solla Price, Big Science, Little Science, p.1.

11. "This Manual deals only with the measurement of research and experimental development (comprising basic research, applied research and experimental development)." OECD, *Frascati Manual*, §17, 17.

international organization specializing in RDI studies. This is not just social technoscience but also an example of a technoscientific agency specializing in the social sciences and, more specifically, in economics and political science of science: the OECD. Having underscored this point, one should also point out that the authors of the *Frascati Manual* were also clear right from the outset that the "I" in RDI embraced activities that were far more wide-ranging than just R&D alone: "besides R&D, other forms of innovative activities may be distinguished in the innovation process."[12] R&D might be at the root of innovation processes and might also have an important function to play through such a process, for example in testing specific new hypotheses that may have merged, but whatever the case, it is a part of the innovation process as a whole. And that is ultimately why the OECD published two manuals instead of just one.

Focusing now on the first of these manuals, the *Frascati Manual*, I will highlight seven points from the work that seem particularly germane to the questions explored in the current book:

1. "The Manual is essentially designed to measure national R&D efforts and to categorise them in various ways . . . In order to ensure maximum comparability with regular economic or social statistics, these are, as far as possible, based on existing UN classifications."[13] The manual does not just attempt to classify and measure R&D, but also to compare statistics from different countries. Its main objective is to achieve convergence and create an international standard for measuring R&D activities. Moreover, it does not just attempt to compare countries but also R&D sectors.[14] This explains the manual's subheading: *Proposed*

12. Ibid., §22, 17.
13. Ibid., §48, 25 and §45, 23.
14. It is worth pointing out that the *Frascati Manual* leaves out military R&D, apparently because for some countries in the OECD, R&D in defense plays a relatively secondary role. This is despite the fact that for some countries involved in developing significant R&D activity, spending on R&D in defense may amount to half or even more of all public spending on R&D (§ 47, 23). It is worth mentioning this in order to recall that there are important military innovation activities that remain outside public scrutiny.

Standard Practice for Surveys on Research and Experimental Development. In order to standardize the measuring and accounting systems of different states, the OECD, through the *Frascati Manual*, elevated itself to the position of a supranational authority with an acknowledged capacity to establish norms that different countries had to fulfill in order to measure the science and technology they practiced. For that reason, this manual established the basis for how to measure R&D. And it was a similar case with the *Oslo Manual* and innovation. According to my own terminology, I would contend that the *Frascati Manual* and the *Oslo Manual* are *technoscientific* rather than scientific works. They deal with science and use the scientific method to classify, compare, and measure it, but their objective is to transform the practice of measurement on the part of diverse agents and institute common regulations to measure R&D. The main goal of the *Frascati Manual* consists of reaching consensus on and passing judgment on how R&D should be measured. This resolve to transform is an integral feature of technoscience, which resorts to scientific knowledge in order to change something existing, not to just to classify, compare, and measure it.

2. Indicator systems are the basic instrument used in order to ensure that ways of collecting and measuring data are the same (or at least similar) across different countries: "This Manual is devoted to measuring R&D inputs . . . Unfortunately, while indicators of R&D output are clearly needed to complement input statistics, they are far more difficult to define and produce . . . innovation surveys are an attempt to measure outputs and the effects of the innovation process . . . Another option is to use existing data sources."[15] This declaration is honest enough to highlight some of the insufficiencies in the methodology used. It ex-

Annex 10 of the *Frascati Manual*, 244–54. addresses the subject, but only in order to classify military R&D activities, not to measure them, because this data either cannot or can only be compared partially on an international level, and not in any detail.

15. Ibid., §14, 17.

plicitly acknowledges that just measuring R&D inputs (investment, resources used, and so on) is insufficient because the results obtained with such inputs should also be measured.[16] However, because this second task is very difficult, the decision is made to survey qualified people (experts) and ask them how to evaluate the results obtained by R&D departments. A methodology like this could be criticized for various reasons due to the fact that it ignores what actually happens and focuses on what those in charge of institutions and companies think about the matter. Inputs can be measured objectively, but outputs have a clear subjective and evaluative component because it is the very people who are in charge of R&D departments who evaluate the results obtained positively or negatively. For that reason, there are two possible ways for public R&D policies or R&D strategies in companies to achieve better results: invest more, with the result that inputs are improved and one country overtakes the others, or try to change the way that experts perceive the results of R&D activities. Opinion about what happens is very important, more so, in fact, than what may have actually happened. Economic investments in R&D (inputs) are real and objective because they can be quantified in monetary terms and are entered into the accounts of government budgets or corporations' investment and spending strategies. Yet in order to know if the results of R&D have been good, average, or bad, one must resort to another methodology based on surveying bosses and experts. Inputs are measured with metrical concepts that offer objective data. However, outputs are quantified via estimates and evaluations that, at most, are based on qualitative and comparative concepts. This is not the classic empiricism of modern science but, rather, a *technoempiricism* that tends to pass judgment on how the data should be interpreted, data that come from indicators rather than real measurement systems. One part of these indicators measures the perception of R&D systems rather

16. The *Oslo Manual* says the same thing about investment in innovation.

than the systems themselves. Even so, because surveys offer quantifiable results, the *Frascati Manual* partially satisfies the scientific requisite given that some of its inputs and outputs can be quantified. In short, there are significant methodological and epistemological problems implied by the simultaneous use of objective and subjective data.

3. Once it has defined its objectives and mentioned its methodological limitations, the *Frascati Manual* defines R&D in very generic terms as, "the creative work undertaken on a systematic basis in order to increase the stock of knowledge, including knowledge of man, culture and society, and the use of this stock of knowledge to devise new applications."[17] The main problem with this definition is that it is reductionist, as is obvious when it distinguishes three kinds of R&D activities: basic research, applied research, and experimental development:

   Basic research is experimental or theoretical work undertaken primarily to acquire new knowledge of the underlying foundation of phenomena and observable facts, without any particular application or use in view. *Applied research* is also original investigation undertaken in order to acquire new knowledge. It is, however, directed primarily towards a specific practical aim or objective. *Experimental development* is systematic work, drawing on existing knowledge gained from research and/or practical experience, which is directed to producing new materials, products or devices, to installing new processes, systems and services, or to improving substantially those already produced or installed.[18]

One could say a lot about these three definitions. I will limit myself to pointing out their strict dependency on the acquisition of knowledge, understanding this only as scientific and technological knowledge, although the latter is not stated explicitly. The arts, for example, remain outside the conceptual framework of the *Frascati Manual*, despite the fact that music, architecture,

17. OECD, *Frascati Manual*, §63, 30.
18. Ibid., §64, 30.

painting, and the other arts require quite a significant amount of knowledge. One could say the same for other forms of knowledge such as philosophical, religious, and juridical. Finally, nor are the different forms of common knowledge, such as that expressed in the natural languages, objects of study according to the methodology of the *Frascati Manual*. Indeed, it focuses solely and exclusively on scientific and engineering knowledge. Consequently, I would conclude that the *Frascati Manual* uses a restrictive and reductionist notion of knowledge. Many creative activities remain outside its field of R&D measurement, with all the consequences that implies as regards policies for fomenting such activities. The *Frascati Manual* is, above all, technoscientific because it is only concerned with scientific and technological knowledge and because, moreover, it establishes how to collect and interpret data, as if they were empirical facts, even when much of this data (the outputs) may have evaluative elements, as demonstrated above.

4. Although basic definitions are drawn up at the beginning, the *Frascati Manual* then enters into a very complex casuistry in order to elucidate what R&D is and is not. It clearly states that neither scientific and technical education and training nor the associated scientific and technological services (for example, libraries, museums, translation, publishing, surveying, prospecting, testing, advisory services, and so on) make up R&D.[19] Curiously, much of what humanist culture used to do remains outside R&D. It is not surprising, then, that later the *Frascati Manual* states emphatically, "No meaningful examples of experimental development have been found in the humanities."[20] It would therefore seem according to the manual that painters, musicians, and artists do not carry out experiments when they draw up sketch outlines, models, or draft versions of their work. This tendency to exclude the humanities from the field of R&D is also evident when it comes to the social

19. Ibid., §19, 18.
20. Ibid., §254, 85. However, different arts incorporate strong experimental components and can be interpreted as R&D in the strict sense of the term: for example, music, paining, and literature itself.

sciences, although to less of a degree.[21] For my part, in chapter 4 I will contrast this R&D model with the C&D (creativity and experimental development) model, which is much wider and inclusive than the conceptual framework used by the authors of the *Frascati Manual*. Numerous other socially important activities are excluded from R&D in the *Frascati Manual*. These include administration and other activities associated with counseling, help, and advice as well as some scientific and technological activities that are similar to R&D and quite a few associated industrial activities. At this point, the authors of the manual make a surprising statement: "The practical definitions given here are intended solely to exclude these activities from R&D."[22] Even so, the fastidiousness that the authors of the manual employ in explaining case by case what is and what is not R&D, after previously having offered apparently precise definitions, is striking. It seems like they were not even minimally satisfied with their conceptual work by means of definitions. I think that there is a more significant reason for all this at root that consists of the "normative" (rather than "descriptive") nature of these definitions. The *Frascati Manual* is a technoscientific text that points out how things should be done (classifications, measurements, surveys, and so on). And the introduction of the term "standard practice" in the title of the work itself illustrates this perfectly. In principle, these standard practices are reduced to a way of carrying out surveys on R&D. However, in excluding multiple human activities related to knowledge and innovation from R&D, and having been transformed into an international standard by

21. One should recall that in the first two editions of the *Frascati Manual* neither the humanities nor the social sciences are considered part of R&D. Instead, these editions just address the natural and technological sciences. Originally, moreover, only the manufacturing sector was considered, ignoring other economic sectors. Later, the need to include the service sector, in which the contribution of the humanities and social sciences is more significant, led the authors of the *Frascati Manual* to include these as possible fields of R&D.

22. OECD, *Frascati Manual*, §67, 30.

which to compare countries and shape science policies, the influence of the *Frascati Manual* in establishing standard practices has gone much further. In the final analysis, it establishes what R&D should be instead of studying empirically what it is. And the same is true of the *Oslo Manual* as regards innovation.

5. This tendency to regulate R&D systems and not just surveys is evident in different passages in the *Frascati Manual*.[23] For example, if the manual's casuistry does not resolve the difficulties in discerning what R&D actually is then one must return to the definitions: "If a particular activity falls within the definition of R&D, then it is regarded as or attributed to R&D; if not, it is excluded."[24] This is a strange form of defining something and implies a normative definition more typical of a legal document than a scientific text. And at root, judgment is being passed on what R&D should be in the OECD countries. If an activity is not adapted to the definition and the additional casuistry, it will not be measured because it will be excluded from the inputs and the subsequent surveys on the results. For the sake of consistency with what I am saying about technoscience, I will use the term *technodefinitions* to describe those definitions that, appearing to be scientific and discuss facts, are in reality normative or regulatory and imply values; most of the definitions in both the *Frascati Manual* and the *Oslo Manual* are of this type.
6. All of this might encourage much debate because it stems from the distinction between scientific and technoscientific concepts. It is a similar case with the concept of innovation, which has obvious axiological connotations and which is far from having clearly defined boundaries. As regards the *Frascati Manual*, I will cite another passage in which the underlying axiological framework is clearly expressed: "The basic criterion for distinguishing R&D from

23. Sheila Jasanoff's views on the regulatory sciences are very appropriate as regards these issues, although I prefer to term them "regulatory technosciences." See Jasanoff, *Science at the Bar*.
24. OECD, *Frascati Manual*, §109, 41.

related activities is the presence in R&D of an appreciable element of novelty and the resolution of scientific and/or technological uncertainty, i.e. when the solution to a problem is not readily apparent to someone familiar with the basic stock of common knowledge and techniques for the area concerned."[25] Later, supplementary criteria are added to the aforementioned basic criterion. I will not discuss these, although if I were to, I would not be straying much off course from what I am suggesting as regards interpreting the *Frascati Manual*. I will just underscore two things: First, in the passage I have just cited, it is clear that the manual is only concerned with scientific and technological knowledge and not in other forms of knowledge. Second, it is stated emphatically, "One aspect of these criteria is that a particular project may be R&D if undertaken for one reason, but not if carried out for another."[26] To me, this argument seems very illustrative of the conceptual undertone of the *Frascati Manual* because the ultimate criterion used to define whether an activity is R&D or not concerns the intentions and objectives of the agents who develop that activity. In other words, if someone seeks out new knowledge and believes that this knowledge is valuable in itself, this is basic research (R). Yet if someone intends for this knowledge to be applicable to resolving some problem or improving a product or process, this is applied research or experimental development (D). The important thing, then, are the goals or aims established by researchers – what they want to achieve. The concept of underlying rationality is teleological and serves to discern R&D activities from those that, according to the *Frascati Manual*, are not. Later on, I will propose an alternative axiological conception of rationality that offers an explanatory model of more general human actions, including those of R&D. This model will also be valid for innovation in general and for its different forms, including social inno-

25. Ibid., §84, 34.
26. Ibid., §85, 35.

vation.

7. Before moving on to the *Oslo Manual*, I will make some general comments about the *Frascati Manual*. The basic criterion referred to above (in §84 of the manual) is a good example of the spirit of the work as a whole. It involves three key terms: novelty in relation to the existing previous knowledge, problem solving, and some improvement in regard to previous results. Novelty is the first stated evaluative criterion: it demonstrates the profound interconnection between the *Frascati Manual* and the *Oslo Manual*. Fundamentally, we are speaking here about scientific and technological innovations that solve problems raised in science and technology or improve previous results. Reference is made to novelty and not innovation because the latter term is reserved for companies and markets. Nevertheless, the idea is the same.

I previously underscored the importance of data comparability for the OECD. The possibility of comparing data and objects is one of the main characteristics of scientific concepts. For that reason, this is a question of an epistemic value (or scientific value) that must be fulfilled so that a piece of data might be considered scientific. I will conclude this discussion by saying that, in the first analysis, the comparability of data, novelty in knowledge, resolution of problems, and improvement of results make up the axiological nucleus of the technoscientific *Frascati Manual*. Other values that shape the measurement methods advanced in this manual could be mentioned, but some of these are more evident in the *Oslo Manual*, which I will now discuss.

Measuring Innovation: The *Oslo Manual*

The subheading of the second manual in the OECD series is titled "Guidelines for Collecting and Interpreting Innovation Data," and this manual constitutes an instrument for measuring innovation activities in different countries and comparing their respective systems. The OECD and Eurostat, the two bodies that designed and applied this to different countries, consider the *Oslo Manual* to be a continuation of the *Frascati Manual*, given that innovation is a wider concept than that of R&D: "While R&D plays a vital role in the innovation process, much innovation activity is not R&D-based, yet relies on highly skilled workers, on interactions with other firms and public research institutions, and on an organisational structure that is conducive to learning and exploiting."[27] In reference to the manufacturing industries, the *Oslo Manual* recognizes that there are several innovation activities that do not depend on R&D:

> As the *Frascati Manual* states, R&D is only one step in the innovation process. Innovation involves a number of activities not included in R&D, such as later phases of development for preproduction, production and distribution, development activities with a lesser degree of novelty, support activities such as training and market preparation for product innovations, and development and implementation activities for new marketing methods or new organisational methods. In addition, many firms may have innovation activities that do not involve any R&D.[28]

These passages from the *Oslo Manual* are in my opinion very significant, which is why I cite them in detail. They reveal that the threefold RDI, understood in a linear way, only covers a part of all innovative activities. This has important consequences when it comes to designing public innovation policies as well as business strategies because, in order to encourage innovation, it implies that it is not enough to just foment R&D.

Traditionally, innovation policies have been conceived as a continuation of R&D policies: in many OECD countries.[29] I believe that this is one of the reasons for the so-called European paradox, the high level of scientific research in the European Union (EU) alongside a less favorable situation when it comes to innovation, certainly compared to the United States, Japan, and other

27. OECD and Eurostat, *Oslo Manual*, §72, 36.
28. Ibid., §310, 90.
29. As the *Oslo Manual* states, "Innovation policy has developed as an amalgam of science and technology policy and industrial policy. It takes as a given that knowledge in all its forms plays a crucial role in economic progress, and that innovation is a complex and systemic phenomenon." Ibid., §24, 15.

countries. There are very innovative countries and regions that are not at the forefront of scientific research. Is this on account of the fact that innovation policies must be designed on the basis of new criteria and not those that have traditionally shaped R&D policies?

I will leave this question unanswered. I think one of the main strategic errors in innovation policies is the belief that innovation always comes from R&D. In some cases this is true, but insofar as innovation processes grow and diversify, there tends to be less of these cases. I will demonstrate below that there are multiple and diverse sources and agents of innovation, which leads me to advocate a pluralist model of innovation. As regards the point I am discussing, that model demands designing innovation policies for the arts, culture industries, humanities, and public sector, and in order to foment the culture of innovation in civil society – not just among scientific researchers, companies, and R&D departments. Many innovations originate in very different circumstances to those of research laboratories, and come from ordinary people, consumers, clients, and users. In particular, there are "disruptive social innovations" that come out civil society itself.

Of the many examples of this, I will just mention two, both of them very important: First, the struggle of the British suffragettes in the late nineteenth and early twentieth centuries for women's right to vote and the later emergence of a gender equality movement, both of which implied disruptive social innovation and both of which radically transformed labor markets, laws, and social life in most western countries. Second, the environmental movement, whose origins lie in the coming together of a small number of activists in favor of preserving certain landscapes and natural environments; the emergence of this social movement ultimately led to the intense contemporary global debate on climate change that, among other things, is giving rise to important innovations in the energy sector (eco-innovation), as the emergence of renewable energy industries demonstrates. Likewise, people's behavior in regard to recycling waste and similar social habits have also been transformed as a result of these changes in attitude.

Returning to the *Oslo Manual*, I will comment briefly on the conceptual evolution it experienced through its three editions

(1992, 1997, and 2005). The first edition (1992) focuses on technological innovation and, within that, on product and process innovations (which it terms "technological product and process" or TPP). Yet when the first prototype EU surveys on innovation were carried out, it was clear that "much of innovation in service sectors is not adequately captured by the TPP concept;"[30] this, as a result of the fact that innovation is not always technological in nature.[31] The second edition of the manual (1997) is also designed to address innovation in the service sector. However, its empirical use in several studies carried out by the EU, Japan, Canada, Australia, and other countries demonstrates other insufficiencies, which in turn led to a further and more profound conceptual revision of the text: specifically, "it was decided to address the question of non technological innovation in this revision."[32] In the third edition (2005), two new forms of innovation are added to the object of study: organizational and marketing innovation.[33] Likewise, it assumes the systemic conception of innovation that stems from evolutionary economics and that is based on the notion of a national system of innovation. Although the manual acknowledges its initial debt to Schumpeter's theory, it is true that the 2005 edition is based on a plural conceptual framework, as its authors recognize explicitly.[34]

That pluralism is one of the points the manual especially underscores: "Innovation can occur in any sector of the economy, including government services such as health or education. The Manual's guidelines, however, are essentially designed to deal with innovations in the business enterprise sector alone."[35] Similarly, it is accepted that "innovation is also important for the public sector. However, less is known about innovation processes in non-mar-

30. Ibid., "Foreword," 3.

31. "Innovation in services oriented sectors can differ substantially from innovation in many manufacturing-oriented sectors." Ibid., §9, 11.

32. Ibid., "Foreword," 3.

33. One should underscore here the fact that these two forms of innovation are not considered technological. For that reason, it is erroneous to identify social innovation with non-technological innovation, as many observers do by establishing a false dichotomy between technology and society. In fact, almost all technological innovations imply social innovations as well.

34. See OECD and Eurostat, *Oslo Manual*, §17, 14. .

35. Ibid., §27, 16.

ket-oriented sectors. Much work remains to be done to study innovation and develop a framework for the collection of innovation data in the public sector. Such work could form the basis for a separate manual."[36]

In sum, the concept of innovation has been extended considerably in recent years. It is commonly recognized that there are technological and non-technological innovations, and that it is not just companies or R&D departments that innovate. Innovations stems from many sources, as Eric von Hippel contends.[37] Furthermore, this plurality has something to do with scales of innovation, in other words, with the microcosmic, mesocosmic, and macrocosmic nature of the environments in which innovation processes are developed. Without rejecting the methodology put forward in the *Oslo Manual*, which has been used by different countries to measure technological and business innovation, one should expand that methodology to embrace the three aforementioned dimensions (sources, types, and scales), each of which is plural in its very structure.

In order to take some steps in that direction, it would be a good idea to analyze in detail the set of contributions implied by the *Oslo Manual* for innovation studies. Another of this work's merits consists of recognizing and defining precisely its own limits, especially in the following four areas that are clearly marked:

> The Manual covers innovation in the business enterprise sector only.
> It deals with innovation at the level of the firm.
> It covers four types of innovations: product, process, organisational and marketing.
> It covers diffusion up to "new to the firm."[38]

It does not, therefore, cover the emergence of new markets, except when they affect a specific company and lead it to encouraging innovation processes in order to adapt to these new circumstances. Nor does it address social change, much of which is

36. Ibid., §28, 16.
37. See von Hippel, *The Sources of Innovation* and *Democratizing Innovation*.
38. OECD and Eurostat, *Oslo Manual*, §26, 16.

a source of innovation. Nevertheless, precisely because the extent of its study is limited, it uses precise methods and introduces quantitative measurement and comparison tools that could be extended to other innovation forms. Its authors are clear on the point that not all change is innovation: "The Manual deals with changes that involve a significant degree of *novelty* for the firm. It excludes changes that are minor or lack a sufficient degree of novelty."[39] Innovation, therefore, is a question of degree, like any other measurable value. Moreover, originality is not the same as innovation: "However, an innovation does not need to be developed by the firm itself but can be acquired from other firms or institutions through the process of diffusion."[40] This observation is relevant for the present study because it demonstrates that creativity only produces innovations on some occasions: specifically, in order for this to be the case, creative ideas must be implemented (developed) and, in particular, disseminated (diffused).

As regards the latter, it is defined in the following way: "Diffusion is the way in which innovations spread, through market or non-market channels, from their very first implementation to different consumers, countries, regions, sectors, markets and firms. Without diffusion, an innovation has no economic impact."[41] Business innovations are diffused via marketing and commercialization, but other types of innovation are channeled through other circuits. Scientific advances and discoveries have their own channels of diffusion: specialist journals, which make up distinct spaces in which to disseminate knowledge. Technological and business innovations are diffused through markets that, from the perspective of innovation studies, become *means of diffusion*.[42] Thus, paraphrasing all these arguments, I will contend that, without diffusion, innovation (whether technological or social) has no social impact. Interpreting the *Oslo Manual*, there is the suggestion that

39. Ibid., §36, 17. Here and in chapter 3, the *Oslo Manual* distinguishes three types of novelty: "new to the market, new to the world and disruptive innovations."

40. Ibid.

41. Ibid.

42. It goes without saying that advertising and the media are other kinds of means of diffusing innovation, hence the importance of marketing.

in other types of innovation one would also have to distinguish between innovation properly speaking and its social diffusion. And success will depend above all on this second phase, without any detriment to the interest, originality, and creativity that an innovative proposal may have, of whatever type it may be. To my mind, the distinction between invention, innovation, and diffusion is valid for innovation in general, not just for technological and business innovation.

With the aim of analyzing innovation processes in firms, the *Oslo Manual* distinguishes three types of innovative business activities: successful, ongoing, and abandoned.[43] Company managers or bosses award these classifications, which is why the data offered by the manual have an obviously subjective dimension. I believe that if this methodology is valid for technological and business innovation then it can be applied to other forms of innovation, including social innovation. As regards social innovation, some proposals are successful in that they enjoy an initial acceptance and are then diffused in society, others fail, and a third variant has yet to see a definitive evaluation.

From all of this I extract two provisional conclusions: first, that innovation studies must not just address what is successful but should also take into account and examine failure; second, when it comes to carrying out innovation surveys, the important thing is to detect who is endorsing them so that their results may be evaluated qualitatively.

There are many more points of discussion as regards the *Oslo Manual*, but at the philosophical-conceptual level there remains a particularly significant question because it demonstrates a clear evolution of the manual from the linear model to the systemic model. It concerns the selection of relevant agents in an innovation system. Whenever one has to study the relational links of a company with other agents, one must bear in mind the following: government laboratories, universities, policy departments, regulators, competitors, suppliers, and customers.[44] Although some agents may not be mentioned, such as workers and distributors,

43. OECD and Eurostat, *Oslo Manual*, §42, 18.
44. Ibid., §50, 20.

on this point the *Oslo Manual* prefigures the pluralist model I propose for innovation studies, which is why I cite this section from the text. The systemic conception of innovation must be applied to any form of innovation. The final point of the *Oslo Manual* that I wish to highlight concerns section 101, which I quote at length:

Drawing on the discussion of innovation theory above, the concept of innovation used in the Manual refers to changes characterised by the following features:

a) Innovation is associated with *uncertainty* over the outcome of innovation activities. It is not known beforehand what the result of the innovation activities will be, e.g., whether R&D will result in the successful development of a marketable product or how much time and resources will be needed to implement a new production process, marketing or organisational method and how successful they will be.[45]

Bearing in mind the previously mentioned theories of innovation, the concept of innovation used in the manual refers to changes that are defined by the following characteristics: (1) "Innovation is associated with *uncertainty* over the outcome of innovation activities"; (2) "Innovation involves *investment*"; (3) "Innovation is subject to *spillovers*. The benefits of creative innovation are rarely fully appropriated by the inventing firm"; (4) "Innovation involves the *utilisation of new knowledge* or a *new use or combination of existing knowledge*"; and (5) "Innovation aims at improving a firm's performance by gaining a *competitive advantage* (or simply maintaining competitiveness) by shifting the demand curve of the firm's products (e.g., increasing product quality, offering new products or opening up new markets or groups of customers) or a firm's cost curve (e.g., reducing unit costs of production, purchasing, distribution or transaction), or by improving the firm's ability to innovate (e.g., increasing the ability to develop new products or processes or to gain and create new knowledge)."[46]

These considerations demonstrate that business innovation is not an end in itself, but that it is subordinate to a canonical entrepreneurial value: increasing productivity.[47] It is also clear that in-

45. Ibid., §101, 34.
46. Ibid., §101, 34–35.
47. In this regard, the opening statement of the manual is very revealing:

novation is linked to the idea of using new knowledge, and therefore to the appropriate management of knowledge, but with the particularity that, in such cases, knowledge is a means by which to achieve successful innovation rather than an end in itself.[48] This is one of the reasons why I argue the case for a close link between innovation studies and technoscience: for scientists, knowledge is an end in itself; however, for technoscientific firms that invest in innovation, it is just a means by which to improve the economic returns of the company via innovation. Furthermore, it is possible to make use of the innovations of others and find more success with these than their original designers: it is often the case that inventors do not benefit from their inventions and that innovative agents are a third party, such as a company that produces the best technological developments, that best markets for a product or process, or that simply has an extra competitive edge because it is better organized. Business innovation is much more complex than research and invention because it requires technological development, organization, marketing, and diffusion. In contrast to what much of the literature on innovation contends, it is not enough to be creative or inventive but to be innovative: according to the *Oslo Manual*, the success or failure of business innovation is measured in markets, where innovative forms compete with one another. Given that innovative activities imply some financial cost for the companies that undertake them, investment (sometimes considerable) is required in innovation. Put another way, investors also form part of innovation networks or systems, as the case of Silicon Valley demonstrates with its venture capital companies, an essential component of this model.

Is it the same case for social innovation? In principle no, because social innovation is not floated on the stock market. However, I tend to think that social innovation also requires some kind of investment, at the very least as regards time and effort. Moreover, after a certain time social (as well as cultural, artistic, and so on) innovation has an economic impact, and even generates companies

---

"It is widely accepted that innovation is central to the growth of output and productivity." Ibid., §1, 10.

48. "Knowledge is more and more viewed as a central driver of economic growth and innovation." Ibid.

and markets. The *Oslo Manual* underscores the fact that innovation processes are not determinist, nor is success guaranteed: indeed, uncertainty is one of the conditions of innovation, as Gurrutxaga has highlighted on many occasions.[49] The same thing happens in cases of social innovation, whose success depends on multiple factors, some of which are unforeseeable because they depend on the disposition of various people.

From the five aforementioned characteristics, I come to the conclusion that there is an underlying hierarchy of values in the *Oslo Manual* that could be summarized as follows: knowledge is a means in order to innovate, yet innovation is a means by which to increase the productivity of a company, which, in turn, is a means of improving its economic results. The ultimate goal, then, is economic benefit for the company, with knowledge, innovation, and productivity just other means to increase such financial gain. The important point is that these values/means, to use Weber's terminology, are hierarchized. Neither knowledge nor innovation occupy the top of the pyramid of entrepreneurial values, in contrast to what one typically hears in much of the informational discourse on innovation, which sometimes appears to posit innovation as an end in itself. At the end of this work I will return to these axiological issues, but I should point out now, at this point without any great risk of being mistaken, that the conceptual framework underpinning the *Oslo Manual* is based on a value system with a very specific structure and hierarchy, and one that is adapted very well to the dominant values of capitalist market economies.

I end my discussion of the *Oslo Manual* by emphasizing this point once again: innovation activities are more extensive, complex, and varied than R&D activities. The manual itself, which I quote at length, affirms this in the following terms:

The specific innovation activities that firms can use to develop or acquire innovations include R&D and/or many other activities that are summarized below:

> *R&D*: R&D is defined in the *Frascati Manual* (OECD, 2002) and includes the following:
> i) The firm can engage in basic and applied research to ac-

49. See for example Gurrutxaga, "Recorridos de la innovación," 43.

> quire new knowledge and direct research towards specific inventions or modifications of existing techniques.
> ii) It can develop new product or process concepts or other new methods to assess whether they are feasible and viable, a stage which may involve: *a)* development and testing; and *b)* further research to modify designs or technical functions.
> *Other innovative activity*: The firm can engage in many non-R&D activities that can be part of innovation. These activities can strengthen capabilities that enable the development of innovations or the ability to successfully adopt innovations developed by other firms or institutions:
> iii) It can identify new concepts for products, processes, marketing methods or organisational changes: *a)* via its marketing side and relations with users; *b)* via the identification of opportunities for commercialisation resulting from its own or others' basic or strategic research; *c)* via its design and development capabilities; *d)* by monitoring competitors; and *e)* by using consultants.
> iv) It can buy technical information, paying fees or royalties for patented inventions (which usually require research and development work to adapt and modify the invention to its own needs), or buy know-how and skills through engineering, design or other consultancy services.
> v) Human skills can be developed (through internal training) or purchased (by hiring); tacit and informal learning – "learning by doing" – may also be involved.
> vi) It can invest in equipment, software or intermediate inputs that embody the innovative work of others.
> vii) It can reorganise management systems and its overall business activities.
> viii) It can develop new methods of marketing and selling its goods and services.[50]

Section 103 of the manual here enumerates the canonical innovative activities, always regarding companies as the innovative agents par excellence. In order to examine other forms of inno-

50. OECD and Eurostat, *Oslo Manual*, §103, 35–36.

vation, such as social innovation, one would have to proceed in a similar way, distinguishing from the outset the socially innovative agents that carry out such innovation. The *Oslo Manual* analyzes business innovation processes in economic terms (inputs) and then uses surveys, questionnaires, and qualitative methodologies in order to study outputs.[51] If a firm carries out some of those previously enumerated actions, and if it does so on the basis of strategic planning, then it has begun an innovation process and can be considered an innovative firm.[52] It is a question of exploring whether a similar methodology would be applicable to social innovation studies. In order to do this, it would be necessary to define who is a socially innovative agent and not just say what social innovation is. One would also have to distinguish socially innovative activities and, finally, one would have to evaluate their inputs (the human resources involved in them, for example) and outputs. The qualitative methodology of questionnaires and quantitative methodology of surveys would also seem applicable to social innovation processes.

Returning to business innovation, the general reflections presented in the first two chapters of the *Oslo Manual* allow it to formulate a series of definitions in chapter 3 that identify the canonical conceptual framework of innovation studies for the OECD and Eurostat: "An innovation is the implementation of a new or significantly improved product (good or service), or process, a new marketing method, or a new organisational method in business

51. The authors of the manual are conscious of the fact that it is not always possible to measure innovation scientifically: "In practice, only parts of such a system can be covered by indicators, while other parts call for qualitative information. Moreover, as policy and indicator analysts are well aware, indicators only occasionally relate neatly to a single factor or issue; more often than not they relate to a range of matters and only partially to each. Any broad information or monitoring system will need to be supplemented with case studies or specialised surveys when specific, in depth analysis is required." Ibid., §121, 40.

52. "The basic definition of an innovative firm . . . is a firm that has implemented at least one innovation, while a product or process innovator is defined as a firm that has implemented either a product or a process innovation." Ibid., §212, 58.

practices, workplace organisation or external relations."[53] As one can see, the concept of innovation depends basically on two others: "implementation" and something "new." The remaining part of the definition distinguishes the four types of innovation accepted by the manual. And the word "implement" is a technical term defined as follows: "A new or improved product is implemented when it is introduced on the market. New processes, marketing methods or organisational methods are implemented when they are brought into actual use in the firm's operations."[54]

These two passages demonstrate once again that the *Oslo Manual* is primarily concerned with innovation that appears and is diffused in markets and with what happens inside companies. The first quote illustrates that product innovation continues to be at the canonical heart of economic studies of innovation, while the remaining three types of innovation are also focused on companies and markets, although more indirectly and in more of a long-term context. A marketing or organizational innovation that has no positive effect for the company will not be considered successful, and will therefore be abandoned. The subordinate status of the "innovation" value in relation to the "productivity" value means that the whole theory of business innovation revolves around one priority objective: increasing productivity in order to augment economic gains of various kinds (profits, an increased share of the market, stock market capitalization, and so on). Without disregarding other considerations that could be made, the *Oslo Manual* is clearly based on improving the extent to which certain economic values and values are met, for which innovation is just one of several means to do so. Nowadays, the importance of innovation and knowledge is repeatedly stressed, and reference is even made to knowledge societies, if not to knowledge communities[55]. All of this shows a potentially promising tendency, but one should not overlook the fact that, when it comes to measuring innovation processes scientifically, the instrument used is the *Oslo Manual* and those notions derived therein. This manual conceives inno-

53. Ibid., §146, 46.
54. Ibid., §150, 47.
55. See Echeverría, Alonso and Oiarzabal, Knowledge Communities, Introduction.

vation in strictly economic and business terms. For that reason, if one wants to analyze other forms of innovation, one must analyze which values — for example, social cultural, environmental, and so on — stimulate innovative processes. The *Oslo Manual* depends to a large extent on the value of productivity, and that is why its main canonical feature is product innovation. Insofar as innovations are designed to promote, for example, social well-being, the conceptual framework of the *Oslo Manual* needs to be profoundly redefined. It might be the case that, in the event of social innovation, increasing public satisfaction could be transformed into an objective to be served by innovation processes. If this were the case, greater attention would have to be given to public services as well as others that the *Oslo Manual* barely addresses at all, such as leisure, entertainment, culture, and tourism, to mention just four kinds of activities that help the well-being of people. I believe such considerations must be taken into account when elaborating a theory of social innovation, and even more so if one is seeking to build a system of indicators to measure it. Insofar as we may want to encourage some kind of scientific study of social innovation, we would have to employ a specific conceptual framework and an appropriate methodology. For these reasons, in carrying out such an exploration it still seems plausible to follow the *Oslo Manual* as one might a train of thought, at least during the initial stage of any such exploration. Although, the conceptual and axiological framework that underpins the manual would have to be remodeled profoundly. For example, both its excessive economic focus and its exclusive emphasis on companies would have to be put to one side. Many agents can carry out social innovation, not just companies: my proposals diverge completely from those of the *Oslo Manual* on this point, as we will see in the following chapter. Nevertheless, one should not reject the methodology of measurement, surveys, and indicators, at least if we want social innovation studies to form part of the social sciences and technosciences rather than becoming to some degree fashionable ideological discourses on innovation. Distinctions and definitions are indispensable because they allow one to collect empirical data in one or more sections. In this sense, the *Oslo Manual* makes another interesting contribution in that it classifies innovation activities into various categories:[56]

1. R&D activities (in-house within a company or external as a result of being purchased from public or private organizations or from other firms).
2. Activities related to product and process innovation, which in turn are subdivided into: (a) the acquisition of other external knowledge (the rights to use patented and non-patented inventions, knowhow, and so on); (b) the acquisition of machinery, equipment, and other capital goods; (c) other preparations for product and process innovations (design, planning, and testing for new products, production processes, delivery methods, and so on); (d) market preparations for product innovations (market studies and so on); and (e) training, in order to introduce and develop innovations.
3. Activities for marketing and organizational innovations, which include preparations for marketing innovations, acquisitions, and so on.

The studies deriving from OECD and Eurostat surveys are usually quantitative, but there are qualitative examples as well. Because companies are the only innovative agents recognized by the *Oslo Manual*, the methodology consists of asking bosses or managers about the innovation activities they are preparing, developing, or have developed, together with their results, which are evaluated qualitatively. And a methodology like this can be transferred to social innovation studies as well as studies of other kinds of innovation (cultural, artistic, environmental, and so on). Yet the key point is determining who the innovative agents are in these cases and what innovative activities they develop. From this basis one must analyze the results (outputs) depending on the evaluations of the people involved (stakeholders) in, for example, such social, cultural, and artistic innovation processes.

My provisional conclusion in this section is that some of the proposals in the *Oslo Manual* are applicable to the study of other forms of innovation, and specifically to that of social innovation,

56. Adapted from ibid., §351, 97–98.

but in order to do this, the manual's initial theoretical and methodological framework needs to be remodeled significantly.

Problems of the STI Model in Innovation Policies and Studies

In the 1990s, the dominant paradigm in innovation studies focused almost exclusively on innovations generated by companies from a basis of technological developments that were applications of scientific advances. This was the STI (science, technology, and innovation) model, according to which scientific and technological knowledge is the main source of innovation, especially in the case of disruptive innovations such as the ICTs, biotechnologies, and nanotechnologies. For this knowledge to be valuated, companies need to be involved because they transform scientific and technological advances into products that are either successful or unsuccessful in the market. Clarifying this analysis somewhat, the model could be termed STEMI (science, technology, enterprise, market, and innovation), given that innovations are only produced in markets with companies (enterprises) being the main source of such innovations. A key moment here is *knowledge transfer* from universities and research centers to companies, and this is the reason why public policies must prioritize establishing collaborative links between academia and the business sector, as suggested by the triple helix model made up of universities, industry, and government. Indeed, lately, some observers have also begun to study innovations in the public sector,[57] conceding that there may be innovations in public administrations, services, and more generally, the public sector as a whole. The systemic approach of innovation studies implied a rejection of sorts of the linear model, but not the complete abandoning of all its postulates. Bengt-Åke Lundvall introduced the concept of an "innovation system" in 1985, reinforcing this idea some years later in 1992 and finding support in the work of Richard R. Nelson.[58] Meanwhile, in 1995 Christopher Freeman applied the notion of innovation to Japan and used the

57. On this, see chapter 4.
58. See Lundvall, *Product Innovation and User-Producer Interaction*; Lundvall, ed., *National Systems of Innovation*; Nelson, ed., *National Innovation Systems*.

term "national innovation system."[59] Since then, the World Bank, the OECD, Eurostat, and different United Nations institutions have used it as a technical concept, distinguishing innovation systems according to country, region, sector, and local environment. The report *National Innovation Systems* (1997) by the OECD accepts that innovation depends on the fluidity of knowledge among diverse agents whose links, relationships, and interactions should be analyzed.[60] And scientists continue to be one, although not the only, source of innovation: "Innovation is thus the result of a complex interaction between various actors and institutions."[61] For that reason, the STEMI model is no longer the only significant model available to study innovation processes. To my mind, that change should go further still, even to the extent of altering the very letters in the RDI acronym because they still evoke the linear model.

For that reason I propose the polyhedral SIDR model (society, innovation, development, and research) in which each of the vertexes of the tetrahedron has its own forms of interacting with the others. This model includes the STI model (RDI) as one of its facets: in other words, one only need substitute the symbol "S" (society) for that of "E" (enterprise, company, organization). Social innovation, in which innovative proposals originate in civil society itself, also fits the same schema. Furthermore, the three remaining sides of the tetrahedron represent three other possible forms of innovation that must also be taken into account. Technical developments properly implemented by a firm can generate innovations even though they may not be based on the foundation of scientific research. In turn, the symbols "D" and "R" can also be replaced by other components in the innovation systems because it is not always the case that innovations emerge out of technological developments and scientific research.

There are innovation processes that stem from very different sources and that are developed in different ways to that of the STEMI model. In *The Sources of Innovation* (1988), von Hippel demonstrates that users, distributors, and suppliers are also sources of innovation; not just the manufacturers and producers

59. Freeman, "The National System of Innovation in Historical Perpective."
60. OECD, *National Innovation Systems*, 4.
61. Ibid., 12.

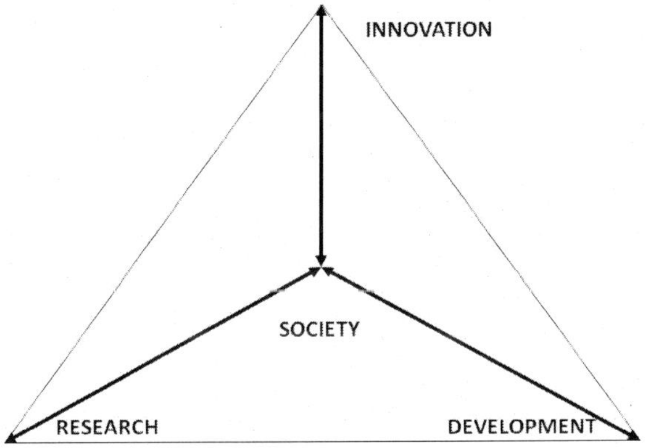

Figure 1. The SIDR Model of Innovation.

of goods and merchandise.[62] In a later work, *Democratizing Innovation* (2005), von Hippel estimates that product and service users produce from 10 to 40percent of innovations that later enjoy widespread social acceptance.[63] Therefore it is not just manufacturers, through their R&D departments, that innovate but also suppliers, distributors, and users, and all this in the manufacturing sector, which is studied by von Hippel and his followers. If we bear in mind that, as the *Oslo Manual* states, innovation is possible in any economic sector, the sources of innovation will differ according to the specific sector. At the same time, the key agents will also differ according to the type of innovation under consideration. The *Oslo Manual* distinguishes seven different types of agent apart from companies: government laboratories, universities, policy departments, regulating authorities, competitors, suppliers, and customers.[64] In the case of social innovation, many of these agents bare-

62. See von Hippel, *The Sources of Innovation*, especially chap. 3.
63. Von Hippel, *Democratizing Innovation*,19.
64. OECD and Eurostat, *Oslo Manual*, §50, 20.

ly have any role in innovation processes. For that reason, much consideration must be given to the potential sources and agents of social innovation, which in general do not coincide with those mentioned in the *Oslo Manual*.

Consider this recent example: a study by the European body Innometrics based on data from the Innobarometer between 2007 and 2009 states, right from the outset, that, "users play an important part in innovation processes and outputs."[65] More specifically, the empirical data show that more than half of European firms that have set up innovation processes involve their users to help these processes. The authors of this study, Stephen H. Flowers, Tanja Sinozic, and Parimal Patel, classify those firms engaged in user innovation as "super innovators," owing to the fact that "compared to other innovative firms, they are more likely to introduce new products, processes and services; they are also more likely to initiate new organizational methods."[66] In sum, according to the data, firms that encourage user innovation lead their rivals that do not in different forms of innovation.

For von Hippel, "distributed innovation" is that which emerges out of plurality of sources of innovation, not just from laboratories or R&D departments.[67] In my opinion, a plural distributed innovation model is indispensable when it comes to analyzing social innovation activities. In contrast to the linear model, which can only be applied to very specific cases of innovation, the pluralist model is a valid means of studying different types of innovation, not just the technological and business variants. The STEMI model is sufficient as a means of studying knowledge transfer processes from laboratories and research centers to companies. However, such processes only generate a part of all innovations. The model I propose, meanwhile, is not reductionist, but pluralist. The main idea behind it is that there is no successful innovation without social acceptance at some point.

Schumpeter and the *Oslo Manual* both distinguish between invention, innovation, and diffusion. And theoreticians of business innovation also contend that the success of any innovation

65. Flowers, Sinozic, and Patel, *Prevalence of User Innovation in the EU*, 3.
66. Ibid.
67. See von Hippel, *The Sources of Innovation* and *Democratizing Innovation*.

depends on its diffusion. Yet they only study diffusion in markets, which is why the *Oslo Manual* only deals with business innovation. Social innovations, in contrast, can be produced in different social settings and not just in markets. Exploring how social innovations are disseminated is one task to be undertaken, and must be done via empirical research and case studies. The SIDR model is not linear but multidirectional, which is why society – and, more specifically, civil society – can be one source of social innovation. I even believe that social innovation that stems from civil society itself constitutes the canon or starting point for social innovation studies. Innovative social culture is demonstrated by addressing the problems and challenges faced by society, which is why non-governmental organizations (NGOs) are good candidates to be acknowledged as potential agents of social innovation. They are not the only socially innovative agents, but by way of comparison, they could be considered the alter ego of companies as regards social innovation. Pluralism in innovation studies implies accepting that there is a plurality of innovative agents, which may be very different according to the different types of innovation under consideration. There are also different areas of diffusion. Another task to be undertaken is detecting those spaces of diffusion and the agents that encourage social innovation within them.

The pluralism I propose for innovation studies does not just refer to agents, sources of innovation, and spaces of diffusion but also to scales and types of innovation. With regard to scales of innovation, the idea is simple: the acceptance of an innovation on the part of society may be demonstrated initially in different areas of varying sizes, without affecting its later diffusion. Some disruptive technological innovations came out of small garages, as the history of Silicon Valley demonstrates. And the same is true of other types of innovation, including social innovation. Attention should be paid to what we might term "small innovations" because some of these have, in time, enjoyed widespread social diffusion. Exploring the diversity of innovation sources implies paying attention to small, medium, and big innovations because innovation processes take place on different scales. As regards types of innovation, it is a subject that, because of its complexity, needs to be addressed in more detail. For my part, I will introduce axiological criteria in order to distinguish different types of innovation. Before doing so, however, it is worth undertaking an initial inquiry into social innovation, properly speaking. The contrast between social, technological, and business innovation will facilitate the distinction between different forms of innovation.

New Innovation Studies in the United Kingdom

Before addressing the new emerging paradigm in innovation studies, I should mention briefly the first alternative proposal to that of the *Oslo Manual*, which emerged in Great Britain during the 2000s. Innovation studies have been developed in various countries in recent years, and in some cases new tendencies have emerged, as we will see later. The EU, for example, publishes its annual *Innovation Scoreboard*, in which there is a classification of the most innovative countries and regions. In order to do so, it has introduced new indicators, some of which are quite significant because they measure innovation without any R&D, something that would have been impossible a decade earlier. However, one thing is methodological changes that correct defects and deficiencies in the dominant paradigm, and quite another is figuring out where and how an alternative paradigm of social innovation emerged. To my mind, this country was Great Britain, even if the contributions of the Crises group from Canada predated this. Here I will briefly discuss some of the changes that have taken place since 2003 in the United Kingdom as regards social innovation studies; this is because I believe certain British institutions have since that time carried out a systematic effort to clarify a conceptual framework for and promote the study of different types of innovation. I was guided in this choice by the fact that, in 2008, the British government published a white paper on innovation, *Innovation Nation*, which implied a significant shift in thinking on innovation in relation to the OECD's *Oslo Manual*, while at the same time basing some of its ideas on the latter text. That white paper marked an important transitional moment for European innovation studies and anticipated the social shift in European innovation policies. In the 2000s, the United Kingdom made a concerted effort to encourage innovation studies, establishing and supporting a series of research agencies, institutions, consultancies, and bureaus in order to develop these studies. As a result of all this, during Prime Minister Tony Blair's final term in office (2005–2007), a plan of action was drawn up regarding innovation that was finally passed and put into practice in late 2008. This plan was drafted and promoted by the UK government Department for Innovation, Universities and Skills (DUS) and presented to the British parliament in March 2008 by the department head, the Secretary of State for Innovation, Universities and Skills John Denham, in the form of a white paper. Prior to its publication, numerous studies had been carried out and published in the United Kingdom, especially by the National Endowment for Science, Technology and Arts (NESTA, now renamed Nesta), but also by the Department of Business, Enterprise and Regulatory Reform (BERR), and the DUS itself, which was responsible for drawing up the annual *Innovation Survey*. Already in 2003, Geoff Mulgan and David Albury were arguing that both government and the public sector could also be innovative agents, and especially the service sector.[68] They therefore broke with one of the main dogmas associated with the linear model, according to which the function of governments and public administrations is to foment business innovation, but without attempting to emulate the private sector. Lord Sainsbury's report, *Science and Innovation Investment Framework 2004–2014*, was also influential because it demonstrated that the British innovation system had its own particular features that were not necessarily included among the more general modish indicators. As the white paper, *Innovation Nation*, states, "The UK also excels at 'hidden innovation' – in its leading services sectors and creative industries."[69] In this way, then, a European government came to officially question the routine OECD and Eurostat indicators, in pointing out that these indicators disregarded various types of innovations. Indeed, some of these overlooked types of innovations were and are vital to the United Kingdom, such as the cultural industries and some services, beginning with different forms of social work that originated in civil com-

68. Mulgan and Albury, *Innovation in the Public Sector*.

69. Department for Innovation, Universities and Skills, *Innovation Nation*, 3.

munities themselves. For its part, the Young Foundation also drew up several historical and empirical studies during the 2000s. These demonstrated that the United Kingdom had been responsible for various disruptive innovations – such as the Open University, a distance learning and research university founded in 1969 with an open entry policy – that did not originate in companies or the government agencies, but rather in the "third sector" or civil society. Thus one of the main dogmas associated with the STEMI linear model came crashing down when it became clear that there were a diverse number of innovative social agents, not just companies. And as a consequence, the British government, without foregoing competing in the field of business innovation, drew up a plan of action in two other directions that were considered equally as important as their business counterpart: "The Government's aim is to make the UK the leading place in the world to be an innovative business, public services or third sector organization; we aim to build an Innovation Nation in which innovation thrives at all levels – individuals, communities and regions."[70]

It is thus stated, and moreover in a government document that defines an entire policy of innovation, that there are several scales or types of innovation and innovative agents, not just technological and business innovations as was previously and widely contended. A program of this kind does not fit the STEMI model. And pluralist and distributed innovation models are required to analyze and interpret it. As well as putting an economic value on things, which typifies successful innovations, there are other forms of adding value, given the fact that there are diverse types of values: for example, social, civic, educational, and environmental values. In short, innovations may stem from any axiological field, and not just in relation to technological, business, or military values.

If anything stands out in the aforementioned 2008 British government white paper, it is the extensive range of documents that were consulted in drawing it up. Allowing for the obvious differences, it could be compared to the 1945 Bush report prepared for President Roosevelt. Just as Bush's *Science, the Endless Frontier* underpinned the beginning of science policy in the United States,

70. Ibid.

so *Innovation Nation* has become the first major reference point for innovation policies in industrially developed countries, precisely due to the fact it broke with the previously dominant paradigm based on the OECD's *Oslo Manual* and, in the final analysis, the linear model as a whole.

Innovation Nation suggests a very general definition of innovation that was previously used by other British institutions such as the Young Foundation and Nesta namely, innovation as "the successful exploitation of new ideas."[71] Immediately, thereafter, though, the white paper clarifies that "new" in this context can refer to an economic sector or an organization in a similar way to that already stated in the *Oslo Manual*. In the manual, the concept of "novelty" refers, in the first place, to companies, although it could also refer to the market as a whole or even the world. *Innovation Nation* points out that a typical form of innovation involves "taking an idea from another context and adapting to another,"[72] in other words, transferring knowledge, adapting it from one situation to another. In order to encourage innovation, therefore, knowledge exchange or trade must be fomented in such a way that the knowledge economy grows and a knowledge society is developed. In the same way as the *Oslo Manual*, *Innovation Nation* accepts that "innovation happens in all service and manufacturing sectors and in the public and third sectors."[73]

I would underscore this statement because it goes further than the conceptual framework of the *Oslo Manual*. I previously cited a passage from the manual in which its authors recognize that it would be possible to publish another manual on innovation in the public sector that would complement those created for the business sector. Along the same lines, *Innovation Nation* introduces a new field of study, *the third sector*, which is usually distinguished from the private and public sectors. I believe this third sector should be the focus of special attention for social innovation studies and that is why I contend that much of the canonical social innovation stems, in the first instance, from the third sector and specifically

71. Ibid., 13.
72. Ibid.
73. Ibid.

from civil society.[74]

I do not deny the fact that the private business sector and the public sector can generate social innovation, just as the third sector is capable of producing important technological and business innovation, even social enterprise. However, following the rules of scientific methodology, social innovation studies must begin with the third sector in a similar way to how the business innovation studies of the *Oslo Manual* focus on companies and the private sector. More specifically, the private and public sectors can be a lot of help in diffusing social innovation, although much of the latter originates in the third sector. The two statements I have just underlined constitute an initial attempt to delimit the field of social innovation studies in such a way that I might proceed in an orderly and methodical fashion. It might be the case that social innovation initiatives find support in public institutions or private firms, as happens in the case of the United Nations Economic Commission for Latin America and the Caribbean (ECLAC) program.[75] However, such support is part of the diffusion phase of social innovation more than the innovation stage properly speaking. Later, one should address social innovation ideas that come out of the private sector (social enterprise) and the public sector (social policies, for example), although such studies must be complementary to those that discuss social innovation in the third sector. Social innovation runs the risk of being devoured by political or business interests, and it is most important to stress its specific nature.

---

74. The notion of a "third sector" means different things in different countries. It dates back to Alexis de Tocqueville, but the term is understood very differently, for example, in the United Kingdom and Japan. In the United Kingdom, in 2006 the government established the Office of the Third Sector (OTS) – subsequently renamed the Office for Civil Society – as a division within the Cabinet Office. It is concerned with community and voluntary groups, social enterprises, charitable and cooperative institutions, and mutual benefit societies. As a whole, all these institutions must be nongovernmental, nonprofit, and they must reinvest any financial gains they do make in subsequent social, environmental, or cultural objectives. By contrast, in Japan, the term "third sector" is used to refer to semi-public corporations, with investment from both the public and private sectors.

75. See Rodríguez Herrera and Alvarado Ugarte, *Claves de la innovación social en América Latina y el Caribe*.

Innovation Nation does not offer any classification of different types of innovation, but it does distinguish between innovation processes based on the aim of making economic gain (business innovation) and those that, to cite some examples, seek to "improve healthcare, tackle environmental problems or improve community cohesion."[76] For my part, I interpret the latter aim in axiological terms: community cohesion is an important social value and there are social innovations that facilitate improving such cohesion, increasing the degree of satisfaction in a specific social value. It is a similar case with other types of values and their corresponding forms of innovation. The authors of *Innovation Nation*, however, focus on goals rather than on values. For them, the objectives of innovation processes offer a criterion with which to distinguish different types of innovation. Nevertheless, one thing is objectives and another is results, and the latter is difficult to measure. Objectives and goals can be examined by means of empirical surveys, so long as one has detected who the socially innovative agents are. However, results are not easy to determine, due to the fact that uncertainty is one of the features of innovation processes in general and of social innovation activities in particular. Moreover, socially innovative activities usually entail unexpected consequences. In fact, it is common to encounter social reforms whose end results are the opposite of what was being aimed for, or that go off in different directions to that expected initially. An innovation process that, due to its initial objectives, appears to be completely social may, later on, have significant economic, political, or business consequences, and vice versa. From an analytical perspective, one must distinguish different types of innovation, but that does not mean that they are separate domains. One should approach innovation in its full complexity and from a pluralist perspective, without reducing it to its economic dimension, or to its social aspect alone.

It is worth noting here the example of the ICTs, undoubtedly one of the major disruptive innovations of our era: their business and technological dimensions are undeniable, but the same could

76. Department for Innovation, Universities and Skills, *Innovation Nation*, 14.

be said for their social component. Internet user communities and some info-social movements – for example, those in favor of open access or free software – have played an important role in the emergence of the information society and economy. Innovation studies require pluralist, flexible, and non-reductionist models: there should be no technological or economic determinism, but nor should there be any social or political determinism. Put another way, *governance of innovation is multifactorial and implies a plurality of significant agents*. The aforementioned British distinction between the three sectors offers an appropriate framework for innovation studies because it allows one to distinguish what comes from profitable private firms, what emerges out of public institutions on the basis of specific policies, and what originates in civil society itself. These three forms of innovation may be interconnected, and may even mutually favor and support one another. However, the agents who promote them are different, and generally speaking, their objectives and strategies also differ from one another. For this reason, the study methodology must be different in each of the three sectors.

I believe that *Innovation Nation* constitutes the first example of innovation governance, specifically because it is not based on improvised ideas but rather on numerous and varied prior studies of innovation. In the following chapter I will discuss some of the documents and reports published by Nesta and the Young Foundation, although other important milestones in British studies and policies on innovation since the turn of the millennium could be mentioned.

3
THE SOCIAL TURN IN INNOVATION STUDIES

Innovation Policies in the European Union

In a March 2003 communication titled *Innovation Policy*, the European Commission (EC) defined EU innovation policy within the framework of the Lisbon Agenda.[1] The communication confirms that the 2002 innovation scoreboard demonstrated that "the innovation performance of the Union remains low in comparison with the United States and Japan."[2] The EC also warned of the scarce growth in productivity in Europe owing to insufficient innovation.[3] As a consequence, it pointed out that there was a risk that the Lisbon Agenda's main objective – achieving EU world leadership in the knowledge society – would not be reached.

This risk was confirmed three years later when the EU confirmed that the objective set in 2000 would not be doable by 2010. The EC then designed a new strategy for 2020, which I will discuss below. First, I will highlight some deficiencies in the conception of innovation used at this time by the EC, two of which are clearly present in the aforementioned 2003 communication.

1. The European Commission had previously published two documents on innovation policy, both related to the Lisbon Agenda: the communication *Innovation in a Knowledge-driven Economy* (2000) and the report *Innovation Policy in Europe 2002* (2003).
2. European Commission, *Innovation Policy*, 4.
3. European Commission, *Productivity*.

In first place, innovation does not just emerge as the consequence of scientific research. The linear model prevailed for years in different European countries and regions, and the widespread use of the letters RDI was one result of this: hence, R&D activities and investments were measured according to the norms, methodologies, and indicator systems of the *Frascati Manual*, while innovation processes were analyzed and measured in line with the guidelines of the *Oslo Manual* (first published in 1992). Later, EU innovation policies were inspired by the systemic model, as stated explicitly in the 2003 communication, *Innovation Policy*: "The evolution of the innovation concept – from the linear model having R&D as the starting point to the systemic model in which innovations arise from complex interactions between individuals, organisations and their operating environment – demonstrates that innovation policies must extend their focus beyond the link with research."[4] This statement is important because of its fresh approach, but it does not suggest the linear model has lost its influence on innovation thinking. Indeed, the same document goes on to say that "the exploitation of invention[s] arising out of the research laboratory is an important, and much studied, route to innovation. Research is a major contributor to innovation, generating a flow of technical ideas and continually renewing the pool of technical skills."[5] And Anthony Arundel and Hugo Hollanders, researchers at MERT (Maastricht Economic Research on Innovation and Technology) Institute, even contend that the linear model is absolutely still in force under the guise of the systemic notion of innovation:

> The countless announcements of the death of the science-push or linear model of innovation, based on R&D, and its presumed replacement with "systemic" models using Schumpeterian definitions of innovation are decidedly premature. The science-push model based on R&D is probably the dominant model in use today by both academics and the policy community. Its continued success is partly due to its successful incorporation of many of

4. European Commission, *Innovation Policy*, 4. On the systemic model, see Nelson, ed., *National Innovation Systems*; Lundvall, ed., *National Systems of Innovation*.
5. Ibid., 6.

> the features of modern innovation theory. These include shifting final outputs from patents to market indicators and evaluating the effect of a range of business strategies. The disadvantage is that this model largely ignores innovation that is not based on R&D.[6]

In second place, after 2003 the EU expanded its working concept of innovation, adhering to an evolution at the international level on the part of the United Nations, the United States, the OECD, and Eurostat. This new wider notion of innovation was reflected in the third edition of the *Oslo Manual* (2005), in which four major forms of innovation are identified: product, process, marketing, and organizational innovations. Each of these is defined precisely and studied empirically by means of their corresponding indicator systems.[7] However, just a few years later one can see that this new expanded notion of innovation is still insufficient. As an alternative, the idea of "social innovation" has emerged and I will examine this concept in the present chapter. This turn toward the social implies a complete change of paradigm as well, rather than just a change or modification of the earlier paradigm. Both the 2003 EC communication, *Innovation Policy*, and the third edition of the *Oslo Manual* in 2005 accept the "economism paradigm" that dominated both innovation policies and studies after the 1980s, although they also extend this notion to new forms of innovation. This paradigm is based on two principles: creating value consists of creating economic value, and the agents who carry out this function are firms.[8] A wholesale acceptance of the economism paradigm is also quite clear in *Innovation Policy*, at least as regards the second of these principles: "Since it is through enterprises that the economic benefit of the successful exploitation of novelty is captured, the enterprise is at the heart of the innovation process. Innovation policy must have its ultimate effect on enterprises: their behaviour, capabilities, and operating environment."[9]

The success or failure of technological innovations is evident in

6. Arundel and Hollanders, 2006 European Scoreboard Methodology Report, 3.
7. OECD and Eurostat, *Oslo Manual*, chap. 3, 47–52.
8. See ibid., §27, 16.
9. European Commission, *Innovation Policy*, 5.

the markets, for example if a firm offers a product at a lower price than its competitors. However, a firm can be more competitive if it innovates and offers new products or services, which reactivate the markets. It can also be more competitive if it improves its advertising or marketing (marketing innovation) or encourages organizational change that reduces costs or improves productivity (organizational innovation). Whatever the case, innovation is a means by which firms can increase productivity and competitiveness.[10] The EC also fully accepts this dominant paradigm: "It is the enterprise that organises the creation of value."[11] Yet this does not stop it from acknowledging that innovation also has a social dimension. Indeed, the definition of innovation for the EC in 2003 was "the successful production, assimilation and exploitation of novelty in the economic and social spheres."[12] Looking at this definition, then, the door has been left open to a social turn in innovation policies, even though that threshold was never subsequently crossed. In fact, the concept of "social innovation" does not even appear in official EC documents. The dominant paradigm, originating in Schumpeterian thinking, conceived of society as a space in which innovations were diffused, but in no way as a potentially innovative agent. According to this paradigm, companies innovated and society was an area from which to capture customers and gain their loyalty while in tough competition with other firms. Many innovative proposals are not successful in markets because they do not achieve a sufficient wide level of diffusion and social acceptance. The social context is important in both the linear model and the systemic model of innovation. But that does not prevent the aforementioned second principle of the dominant paradigm from still being in force, namely, that firms are the innovative agents par excellence, if not the only ones. Social diffusion raises questions of commercialization, marketing, and competitiveness in markets, all issues that companies must anticipate and resolve if they want to be successful with their innovations. This is one of the reasons why, since Schumpeter, innovation has been distinguished from invention. The former is a process that takes a particular route and

10. OECD and Eurostat, *Oslo Manual*, §1, 10.
11. European Commission, *Innovation Policy*, 6.
12. Ibid., 5.

goes through different phases. If the OECD, Eurostat, and the EC extended the concept of innovation to include marketing innovation as a fourth form, it was precisely because of the importance of diffusion and social acceptance for innovative proposals. Nevertheless, they did not abandon the basic principle associated with the dominant paradigm: namely, that innovative agents are companies.

Innovation Policies in the Revised Lisbon Agenda

When it was clear that it would not achieve the objectives devised for the Lisbon Agenda in 2000, the EU remodeled its entire strategy in 2005, including its innovation policies.[13] The Aho report (2006) played an important role in this regard, in that it puts forward the goal of an innovative Europe, which goes beyond the purely business field.[14] This report introduces the notion of a paradigm shift in innovation policies, arguing that the RDI model is

---

13. The new EU social agenda for the period 2006–2010 is defined in the European Commission communication, *On the Social Agenda*. As regards reworked innovation policies, possibly the most important document is another European Commission communication, More *Research and Innovation: Investment for Growth and Employment*. In the latter document, nineteen priority areas are defined, and strategic actions, such as the creation of initiatives like PRO INNO and Europe INNOVA, are promoted. At that time, both initiatives were focused on entrepreneurial innovation. Another essential document is European Commission, *Putting Knowledge into Practice*, which outlines how the recommendations of the Aho report can be put into practice.

14. Aho, *Report of the Independent Expert Group on R+D and Innovation*. The report is based on the systemic notion of innovation and proposes a "pact for research and innovation" (chapter 5) among political, business, and social leaders. It was drawn up within the framework of the revised Lisbon Agenda, and its idea to promote a culture of innovation, fomenting government action in questions of legislation, regulations, public contracts, and intellectual property was groundbreaking. The second main objective of the report is still R&D as a way of generating innovation in industry, and at its core is an assessment of European R&D policy. The third and final general objective of the report is to foment mobility, both in the field of finances and in that of organizations and knowledge, thereby encouraging knowledge transfer and the creation of business clusters. As such, business clusters have become the innovative agent par excellence in the EU, as a consequence of the Aho report.

too restricted.[15] This is correct, and the fact that a report at this level acknowledges it as such is shows that cracks were already starting to appear in the dominant paradigm, despite the recent corrections to the latest edition of the *Oslo Manual.* Nevertheless, the Aho report limited itself to updating rather than replacing that paradigm, highlighting "the need for modernising the definitions of industrial R&D to reflect the present reality of business research and innovation practice."[16] Expanding on this idea in a footnote, the authors of the report contend the following:

> In particular, it is unacceptable that the current state aid rules for R&D are based on the linear model of innovation when there is a broad academic, gubernamental and industry consensus that the model is wrong. The interactive innovation process that is necessary to take advantage of the market-driven approach we are advocating means that the obsolete, artificial distinction between "industrial research" and "precompetitive development" should be replaced by a single category "industrial R&D" (including prototyping, software, testing and trials) with an allowable aid intensity of at least 50% should be implemented.[17]

As one can see, according to the report, a change of paradigm is called for because the linear model is no longer valid, but it does not propose an alternative paradigm. The priority of R&D in innovation policies is still stressed, even if the model is reworked slightly. And despite its criticism of the linear model, at its core it maintains the same conceptual framework (competitiveness and innovation based on knowledge) stated in the previous EC document, *More Research and Innovation – Investment for Growth and Employment* (2005). When it analyzes European data and compares them to those of the United States and Japan, the Aho report argues above all that European firms invest less than their American and Japanese counterparts in R&D.[18] Yet when the report discusses innovation, it uses the indicators of the Trend Chart European Innovation Scoreboard (2005), which were designed ac-

15. "The paradigm shift cannot be confined to the narrow domain of R&D and innovation policy, important though that is." Ibid., 1.
16. Ibid., 16.
17. Ibid., 16n27.
18. Ibid., 2.

cording to the classic business model.[19] The Aho report once more directed European innovation policy toward R&D, without taking serious consideration of other forms of innovation. It is true that the report marked some significant conceptual advances, for example in accepting that users – and especially "lead users"[20] – can generate innovation, thereby embracing von Hippel's notion of distributed innovation.[21] However, the sectors considered priorities by the report – e-health, pharmaceuticals, transport, logistics, digital content, energy, security, and environment[22] – are typically industrial, with the exception of the environment; although in this latter case, too, the strategy proposed is clearly business-oriented and based on R&D. Indeed, the report even identifies "eco-innovation" with environmental technologies.[23] In sum, the Aho report remains strictly within the Schumpeterian framework of technological and industrial innovation, attempting to modernize and adapt it to new innovative practices, but without straying from the business and industrial sphere.

The Aho report led to certain changes in European innovation policy as was obvious from the EC communication that implemented it in 2006, extending the previous strategy:

> The EU can only become comprehensively innovative if all actors become involved and in particular if there is market demand for innovative products. This broad strategy needs to engage all

19. The European Trend Chart on Innovation was created in 1998 and from January 2000 on assumed the task of elaborating the European Innovation Scoreboard (EIS), which offers comparative data on innovation in different European states. From 2002–2003 on, it also began drawing up the European Regional Innovation Scoreboard (RIS), which deals with the regions of Europe. In 2007, the Trend Chart on Innovation was replaced by the Inno-Policy Trend Chart. As well as data, this published documents on methodology through which one can follow the changes in indicator systems. In one such document, the *Regional Innovation Scoreboard 2009*, its authors Hollanders, Tarantola, and Loschky introduce a particularly important change by including among the indicator systems one that is dedicated to innovation not based on R&D as well as other indicators for non-technological innovation.
20. Aho, *Report of the Independent Expert Group on R+D and Innovation*, 5.
21. See von Hippel, *Democratizing Innovation*.
22. Aho, *Report of the Independent Expert Group on R+D and Innovation*, 8.
23. Ibid., 10.

> parties – business, public sector and consumers. This is because the innovation process involves not only the business sector, but also public authorities at national, regional and local level, civil society organisations, trade unions and consumers. Such a wide partnership for innovation will create a virtuous circle, where supply of new ideas and demand for new solutions both push and pull innovation.[24]

Firms and business are, then, the driving force of innovation, but other social agents can also be involved in encouraging it so long as they do so at a national, regional, or local level. People are thought of as consumers and, at most, as the demand for innovation, in line with the dominant economism model according to which innovation is carried out in markets. However, one does see the beginnings of a mention for other agents and even an allusion to civil society organizations and labor unions as agents in an innovation system. The EC communication *Putting Knowledge into Practice* goes so far as to assert that "all forms of innovation need to be promoted, for innovation comes in many forms others than technological innovation, including organisational innovation and innovation in services."[25] However, it does no go so far as to mention social innovation, and it remains within the conceptual framework of the OECD and Eurostat's *Oslo Manual*. It seeks to promote a culture of innovation and attempts to make European society innovation friendly, but it continues to state that innovation is done by companies and that its social success or failure is explained in market terms.

As regards the instruments used to measure innovation processes, the EU modified the EIS in some important aspects, but without breaking with the former paradigm. The September 2008 Innometrics report, authored by Hugo Hollanders and Adriana van Cruysen, incorporates these changes into the indicators, but for the period 2008–2010 of the EIS.[26] The indicators were now extended to twenty-nine and distributed among seven groups so that, since the EIS was established, only thirteen stable indicators

24. European Commission, *Putting Knowledge into Practice*, 3–4.
25. Ibid., 4.
26. See Hollanders and van Cruysen, *Rethinking the European Innovation Scoreboard*.

have been maintained. This gives some idea of the unstable nature of and the changes experienced by European innovation studies, almost always because of criticism that innovation indicator systems have received.[27] In the same document, an innovation indicator not based on R&D is introduced for the first time and there is partial acceptance of the Chesbourgh (open innovation) and von Hippel (user innovation) models.[28] Moreover, empirical studies on other forms of innovation are carried out for the first time, on the basis of data from the Innobarometer. There was, at the same time, a strong push on the part of the EU to establish industrial clusters, especially at the regional level (termed "innovative clusters") and in line with the open innovation model. Furthermore, the novel concept of eco-innovation[29] was encouraged, but only from a technological and business perspective examined in the Aho report. And empirical studies of user-based innovation were also undertaken for the first time,[30] thereby heading in the direction of social innovation, but without quite arriving there. In sum, these developments called into question the previously dominant paradigm in more than one aspect, yet they did not lead to the establishment of an alternative paradigm, for all the talk of a changing paradigm and a crisis in the linear model. Indeed, all the qualifications and extensions of the innovation model during this time did not hin-

---

27. Annex 2 in Hollanders and van Cruysen, *Rethinking the European Innovation Scoreboard* (35–40) summarizes criticism of the EIS, which falls into five categories: high-tech bias, use of inadequate indicators (and a lack of other indicators), problems with the quality of data from different countries, optimization problems, and statistical problems (correlated indicators). As a whole, both this set of critiques and the periodic changes in indicators demonstrate that the most typically used tools to measure innovation on the basis of the economics paradigm present significant problems.
28. See Chesbrough, *Open Innovation*; von Hippel, *Democratizing Innovation*.
29. European Commission, *Putting Knowledge into Practice*, 13.
30. Of particular interest in this regard is the report by Flowers, Sinozic, and Patel, *Prevalence of User Innovation in the EU*, which argues on the grounds of empirical data that, of those European firms that develop product or process innovation (around 30 percent of the total), more than half of these involve their users in their innovative activities. Such data, being highly reliable, seriously question the linear model and diminish the importance of R&D-based innovation.

der the fact that the new 2007–2013 EU framework model was titled the Competitiveness and Innovation Programme (CIP).[31] The main goal of the revised Lisbon Agenda thus continued to be that of its original incarnation in 2000: to make the EU more competitive in economic terms, especially in comparison to the United States, Japan, and several emerging countries such as China, with innovation conceived as a means to this end. As a consequence, the economism paradigm in innovation is still in force, even though it has been expanded and modified on certain points.

The Emergence of the Social Innovation Paradigm in Canada and the United Kingdom

In contrast to the dominant tendencies of the OECD, Eurostat, and the EC during the new millennium, in the early 2000s certain countries began to formulate an alternative paradigm. Three such countries – Canada, Australia, and New Zealand – are outside the EU, yet the other, Great Britain, is an EU member state. In this section I will explore the social turn in innovation studies (and policies) that took place in Canada and the United Kingdom because these are the two countries that have done most to encourage social innovation as an alternative paradigm to technological innovation.[32]

Canada was one of the pioneer countries in the study of social innovation. In December 1997, the Quebec government designed an innovation policy and, among other actions, commissioned the Conseil de la science et de la technologie (CST, Science and Technology Council) to draw up a study on the potential contribution of the social sciences and humanities to a planned innovation policy. A specific working group on social innovation was then

31. Decision no. 1639/2006/EC of the European Parliament and of the Council of 24 October 2006 establishing a Competitiveness and Innovation Framework Programme (2007 to 2013), published in the *Official Journal of the European Union* 310 (November 9, 2006), 15–40.

32. Benoît Godin demonstrates that the term "social innovation" had already been used in the nineteenth century to refer to socially innovative thinkers such as Charles Fourier, Henri de Saint-Simon, and others. See Godin, *Social Innovation*, 10. However, the term never came to make up a technoscientific paradigm and nor did it generate any reflection on social innovation processes in terms of their structure and phases.

created—the Groupe de Travail sur l'Innovation Sociale (GTIS, the Social Innovation Working Group)—as part of the Conseil québécois de la recherche sociale (CQRS, Quebec Social Research Council), and it compiled an initial document titled *Recherche en sciences humaines et sociales et innovations sociales: Contribution à une politique de l'immatériel* (Research in the humanities and social sciences: Contribution to an intangible policy, 1999).[33] Later, in early 2000, the final CST report was published, titled *Innovation sociale et innovation technologique: L'apport de la recherche en sciences sociales et humaines* (Social innovation and technological innovation: The contribution of research in the social sciences and humanities).[34] Both documents are pioneer works in social innovation studies and contribute the first serious reflection on the issue in that they call attention to "socially innovative practices," which they term "social innovations."

This emerging new paradigm was still, however, diffuse, although it did point out a first distinctive criterion: social innovations confront social problems and not just business, technological, market, or production issues. Yet the GTIS document does not question the linear model because it takes as read that social innovation originates in scientific research in the social sciences, for example in studies of new services or new forms of organization.

A further document was then published in November 2003 that might now be considered as an essential reference point for social innovation studies in the twenty-first century: *Qu'est-ce que l'innovation sociale?* (What is social innovation?) by Julie Cloutier,[35] a member of the Crises group (Centre de Recherche sur les innovations sociales, Social Innovation Research Center) at the University of Quebec, Montreal. In the work, she carries out a very careful examination of what could be termed the "prehistory of social innovation studies," which she dates back to a 1970 article by James B. Taylor: "Introducing the Social Innovation."[36]

33. Gouvernement de Québec, Conseil québécois de la recherche sociale, *Recherche en sciences humaines et sociales et innovations sociales*.

34. Gouvernement de Quebéc, Conseil de la science et de la technologie, *Innovation sociale et innovation technologique*.

35. Cloutier, *Qu'est-ce que l'innovation sociale?*

36. Taylor, "Introducing the Social Innovation." The Crises group also

Cloutier undertakes a detailed revision of the existing bibliography and discusses different definitions and characterizations of social innovation advanced by academics during the late 1990s, but all of which never came to influence the innovation policies then being promoted and, moreover, all of which were strictly technological and business-oriented in scope. In Canada, however, the above-mentioned GTIS and CST reports *were* influential when it came to the Quebec government's innovation policies. For this reason, in my opinion this marks the moment of a technoscientific paradigm shift in innovation studies.

Especially incisive in all this was the Goldenberg report, published in 2004 by the Canadian Policy Research Networks (CPRN). Even the title of the report itself—*Social Innovation in Canada – How the Non-profit Sector Serves Canadians… and How It Can Serve Them Better*—points to a new understanding of innovation.[37] Here, Goldenberg is interested in innovation processes that originate in civil society and, more specifically, at the community level. According to him, social innovation stems above all from the nonprofit sector; this assertion has contributed to establishing a whole new and alternative paradigm because it shows that, as well as companies that seek economic gain, there are other innovative agents in social sectors that are not concerned with business. A couple of years later, this idea was embraced by the Young Foundation in Great Britain through Geoff Mulgan's influential *Social Innovation* (2007), in which he makes a clear distinction between "social innovation" and "business innovation."[38] Mulgan offers two definitions of social innovation: a general notion stating that "social innovation refers to new ideas that work in meeting social goals," and a more specific idea whereby it is classified as "innovative activities and services that are motivated by the goal of meet-

published another important document in Petitclerc, *Rapport sur les innovations sociales et les transformations sociales* (Report on social innovation and social transformation). Here, Petitclerc links social innovation studies to sociological theories on social change. All the studies mentioned have been recently refined by Godin's already noted *Social Innovation.*

37. See Goldenberg, *Social Innovation in Canada.*

38. Mulgan et al., *Social Innovation*, 8. Mulgan originally presented the work in Beijing, China, in 2006, and it was published a year later.

ing a social need and that are predominantly developed and diffused through organizations whose primary purposes are social."[39] Therefore, innovation is social if, in the first instance, the agents that promote it are organizations with social goals. Mulgan and his colleagues at the Young Foundation – Simon Tucker, Rushanara Ali, and Ben Sanders – perhaps in order to differentiate their work from that of Goldenberg (who they do not mention), argue from the outset that, "social innovation is not unique to the non-profit sector."[40] This new orientation in innovation studies was adopted by the British independent charity Nesta (renamed from the original NESTA), an institution that published numerous studies on innovation between 2006 and 2010, including in the fields of social, cultural, and artistic innovation.

Mulgan's *Social Innovation* was an important theoretical aid for the political strategy of British Prime Minister Tony Blair when he established the Cabinet Office of the Third Sector in 2006, which in turn promoted the Social Enterprise Action Plan that same year – one of the major reference points for social innovation policies. British policy in innovation matters culminated in 2008 with the publication of the aforementioned *Innovation Nation*, which as we have seen presents a systematic plan to transform the United Kingdom into a world reference point in innovation policy, including that of social innovation.[41] The Goldenberg model was, then, influential in Europe, not just in the United Kingdom but also in Sweden. There, the Vinnova Foundation commissioned a detailed report on Canadian innovation policies with the aim of using them as an example for implementing similar policies in Sweden.[42]

39. Ibid.
40. Ibid., 4.
41. See Department for Innovation, Universities and Skills, *Innovation Nation*.
42. See Liljemark, *Innovation Policy in Canada*. As regards innovation policies in Sweden, which I will not address in this work, the key text is Ministry of Industry, Employment and Communications and the Ministry of Education (Swedish Government), *Innovative Sweden*. This report does not allude to social innovation, although it does contain a section titled "Innovative People" (40–44) that is clearly oriented toward promoting entrepreneurship, a strategy run by the Swedish Business Development Agency NUTEK.

Goldenberg himself, together with Wathira Kamoji, Larry Orton, and Michael Williamson, updated his 2004 findings with a new report published in 2009 that analyzes in depth the development of the social innovation concept, which was in effect stunning, in the five years since his original publication.[43] The Canadian government did not follow Goldenberg's advice in 2004 but did so after publication of the 2009 report, finally giving official approval to promote social innovation policies. Then President Barack Obama followed suit in the United States, creating the Office of Social Innovation and Civic Participation within the White House in late 2009. This office administers the Social Innovation Fund, with an annual budget of $50 million until 2015. As well as Great Britain, Canada, and the United States, both Australia and New Zealand have also encouraged public policies to promote social and not just technological innovation.[44] I would conclude, therefore, that the social turn in innovation policies occurred around 2004 as a result of the above-mentioned Canadian reports (by the Crises group and Goldenberg) and subsequent British studies (by the Young Foundation and Nesta). It goes without saying that these new innovation policies gave rise to the establishment of different social innovation studies centers, and more broadly speaking research on the same topic, in other countries.[45]

This social turn in innovation studies has led to the rapid development of research into social innovation, which is well summarized in Goldenberg et al.'s 2009 report, which states that the basic

---

43. Goldenberg et al., *Social Innovation in Canada: An Update.*

44. In 2001, the Australian Centre for Social Innovation (ACSI) was established. This is, today, the main national center designed to promote social innovation. New Zealand also created its own Centre for Social Innovation and it even enjoys its own investment group, the New Zealand Social Innovation Investment Group.

45. As well as Crises in Quebec and the aforementioned British groups (the Young Foundation and Nesta) – probably the most active groups when it comes to social and cultural innovation – it is also worth mentioning the Toronto, Stanford, Vienna, Australia, and New Zealand centers. In Spain, social innovation centers have been established by the Ramon Llull University (URL) in Barcelona, the Bankinter Foundation in Madrid, and the University of the Basque Country (UPV-EHU), the latter named "Sinnergiak" and founded in July 2011.

direction being taken is the following: "the role of the non-profit sector remains critically important, but social innovation is seen today as a broader concept—one that cuts across the non-profit, for-profit and public sectors and that addresses a wider and more complex range of societal issues."[46] At root, social innovation stems from civil society (the third sector: the cooperative movement, social movements, voluntary work, environmental groups, charity organizations, and so on). Gradually, however, initiatives originating in both the public sector (social affairs programs directed toward innovation, together with educational, service and public administration, urban innovation, and so forth) and the private sector (social enterprise, philanthropy, foundations, corporate social responsibility, and so on) are being added to the main sources of social innovation. Nowadays, then, it is widely accepted that any of the three sectors can generate social innovation, especially when they cooperate with each other.

The Paradigm of Social Innovation in the European Union

How has the EU reacted to this emerging paradigm? As we saw, the 2006 Aho report left the door open to taking new directions in social innovation without actually recommending doing so. The first European social innovation policies were promoted by specific countries such as the UK, Finland, Sweden, and a few others, sometimes through completely different programs and other times as part of more general innovation policies.[47] The work undertaken by Nesta in the UK, through its publication of numerous studies and reports on different forms of innovation, stands out. Indeed, in my opinion, Nesta currently leads the field in carrying out innovation studies at a European, and even global, level. Of particular interest are its reflections on "hidden innovation," a very important concept that I will address below.

46. Goldenberg et al., *Social Innovation in Canada: An Update*, v.
47. Taipale, ed., *100 sosiaalista innovaatiota Suomesta* (100 innovations from Finland). This book concerns social innovation in Finland but I have not been able to use it because, to date, it has only been translated into Japanese. The Swedish Innovation Agency, Vinnova, finances the Designing Social Innovation program that began in April 2010 and lasted for a period of three years. It was directed by Ramia Mazé.

The revision by the EU of its 2000 Lisbon Agenda, imparting it with a stronger social and cultural stimulus, brought with it a certain social turn in European innovation studies. Yet this change in direction has scarcely been reflected in European innovation policies, except in certain specific countries. I believe, though, that before very long the EU will fully embrace this social turn and integrate it into its innovation policies, seeing the drive toward this objective in the United States, Canada, and elsewhere.[48] In fact, the concept of social innovation is increasingly used in academic research carried out in the developed countries, as well as in public policies and business strategies. Concepts from the same family – corporate social responsibility, social marketing, social capital, social economy, social entrepreneurship, and so on – are now used quite often, which means that the concept of social innovation runs the risk of becoming a hollow label, or "container concept,"[49] without any accurate or precise definition. The World Bank and the OECD, for example, have embraced as their own the concept of social capital. And for its part, UNESCO promotes a social innovation initiative, the ECLAC program, in Latin America and the Caribbean, together with sponsorship by the W.K. Kellogg Foundation.[50] Moreover, the 2009 World Economic Forum Annual Meeting, organized by the Schwab Foundation and held in Davos, Switzerland, was especially important. It included a session titled "Social Innovation in a Post-Crisis World."[51]

As regards specific studies on social innovation, I will just comment briefly on the new directions proposed by Nesta and the Young Foundation because, as noted above, I believe both these institutions are at the forefront of the clear social turn taken in

---

48. In early 2009, the president of the European Commission, José Manuel Barroso, announced the creation of a group of experts on social innovation with the aim of defining the EU position at the World Economic Forum in Davos, Switzerland, at which the topic of social innovation had been included on the agenda. This is the document I will discuss below.

49. As argued by Gurrutxaga and Echeverría, *La luz de la luciérnaga*, 92–95.

50. See www.eclac.org/dds/innovacionsocial/portada_i.htm and Rodríguez Herrera and Alvarado Ugarte, *Claves de la innovación social en América Latina y el Caribe*, in which there is a summary of the first Social Innovation Fairs organized by ECLAC in Latin America.

51. See "Social Innovation in a Post-Crisis World."

European research on social innovation.

In 2006 and 2007, respectively, Nesta published two important reports on innovation: the policy briefing *The Innovation Gap* (October 2006) and the research report *Hidden Innovation* (June 2007). Both documents advance an important hypothesis: namely, that "science-based innovation represents only one important dimension of innovation."[52] Indeed, in the same text Nesta estimates that only 6 percent of innovation in the British economy is science-based and that many innovation processes are not being measured by the typical indicator systems (the OECD, Eurostat, the *Oslo Manual*, Innometrics, and so forth). There thus emerges the hypothesis of "hidden innovation," defined by Nesta in the following terms: "the innovation activities that are not reflected in traditional indicators such as investments in formal R&D or patents awarded. Despite not being measured, hidden innovation often represents the innovation that matters – the innovation that most directly contributes to real practice and performance of a sector."[53]

As a means of responding to this hypothesis, which is tremendously important for both innovation studies and policies, "UK policymakers have begun to address these new and important dimensions of innovation; in particular, the Department of Trade and Industry (DTI) is investigating broader categories of innovation including innovation in services."[54] In fact, both the DTI and the DIUS (Department of Innovation, Universities and Skills) commissioned Nesta to draw up a new Innovation Index that would detect and measure different types of innovation in the United Kingdom, including those that did not originate in R&D. And significant data were revealed in a pilot version of this work, published in November 2009: for example, Nesta estimates that, of the total British private sector investment in innovation (£133 billion in 2007), over three quarters of this was in hidden innovation and only one quarter in traditional R&D.[55] If such data are

52. NESTA, *Hidden Innovation*, 4.
53. Ibid.
54. Ibid.
55. NESTA, *The Innovation Index*, 4. Note: Both the United Kingdom (officially since 1974) and the United States use the short-scale measurement

confirmed, it would mean that in a scientifically advanced country like the United Kingdom, most private-sector innovation activities are not based on scientific research, contrary to the dominant paradigm. Simultaneously, Nesta also published different reports on the creative and cultural industries as well as on innovation in (among other subjects) art, urbanism, health care, and social work. In short, it came to demonstrate irrefutably that there are many other forms of innovation besides R&D and companies, most of which are not taken into account by the empirical studies of Eurostat, the OECD, Innometrics, and so on, precisely because traditional indicator systems are R&D focused. The previously mentioned British government white paper, *Innovation Nation* (2008), which serves as the current basis of UK innovation policy, took on board Nesta's main hypothesis and began to conceive of innovation as a multidimensional process that, on occasion, is produced in the R&D departments of companies, but which on most occasions does not originate in R&D.

Put in Kuhnian terms, then, we are thus faced with a *significant anomaly in the dominant paradigm* that calls the very paradigm itself into question because much of what happens in innovation processes, perhaps even half, remains outside the measurement systems based on the EIS, the RIS, and others. All of this reinforces the idea of a paradigm shift in international studies on innovation, which began in the United Kingdom and is being extended to other European countries. Seen from another perspective, it may be the case that the very instruments used to measure innovation are not just insufficient, but are actually a significant part of the problem. This is, indeed, what Nesta suggests in *Hidden Innovation*: "Historically, the development of metrics has proven a substantial barrier to the extension of innovation policy beyond science and technology."[56] This is because R&D policies give priority to R&D and finance above all these types of innovation activities, ignoring other multiple and diverse forms. If this were the case, the paradigm shift in innovation studies would not be limited to academic

---

system for the term "billion," meaning 10^9, which is more commonly referred to, following the long-scale measurement scheme, as a "thousand million" in European countries and elsewhere.

56. NESTA, *Hidden Innovation*, 5.

debate but, rather, there would be a widespread belief that innovation policies traditionally based on R&D need to be rethought by overcoming that conceptual and metric limitation. We should remember that, as the European *2006 Trend Chart Methodology Report* states as regards indicators systems and measuring methods, "the Trend Chart project is proactive – the thematic reports and the EIS (European Innovation Scoreboard) are designed to provide indicators and analysis that are relevant to current policy concerns. For example, the selection of EIS indicators for the drivers and characteristics of innovation is partly determined by the policy relevance of these indicators."[57]

Put in my own terms, *innovation studies are not scientific but technoscientific*. They do not just seek to know what innovation processes might be in all their fullness and variety, but also to generate terminology, methods, and indicators that allow one to measure and compare such processes. And all of this is at the service of business strategies or public policies aimed at fomenting innovation and linked to certain interests and values. This is equivalent to saying that we are not faced, in this case, with a paradigm in the Kuhnian sense, but rather with a "technoparadigm" – in other words, with proactive knowledge that is not concerned with the "know why," but with the "know how" and the "how to do." The technosciences are different from the sciences in that they are oriented toward transforming the world around them more than understanding what it is like. In the case of the social technosciences, their aim is to transform societies into a specific direction, converting them into information or knowledge societies, as in the main aim of the 2000 Lisbon Agenda. Following the 2006 Aho report, however, talk has focused on ideas like an "innovative Europe," "innovative societies," and even "innovative nations."[58] Innovation studies are a form of technoscience and are as such proactive. And indicator systems are tools that serve to implement specific active policies, normally with the goal of creating or transforming national, regional, or local innovation systems. Innovation technoparadigms

57. Arundel and Hollanders, *2006 Trend Chart Methodology Report*, 1.
58. As in the case of the aforementioned British publication, *Innovation Nation*, or Innobasque (the Basque Innovation Agency) in the Basque Country.

(the *Oslo Manual*, Union Innovation Scoreboard, and so on) are, above all, know-how systems that determine how the innovation they are interested in should be conceptualized, defined, and measured, and are defined according to such policies even though a lot of innovation remains outside the dominant metrics, as Nesta points out when discussing hidden innovation. For my part, I anticipate that these conceptual and methodological changes in innovation studies are the prelude to changes in the corresponding policies, and that is why I am attempting to assess what the new conceptual frameworks might be.

Thus, British studies on social innovation ended up influencing EU thinking on the matter. In fact, in December 2008 the Council of Europe summoned several expert panels with the aim of drawing up a European Plan for Innovation that would replace previous similar initiatives. In order to take the first steps in that direction, the Bureau of European Policy Advisers (BEPA) organized an innovation workshop in Brussels on January 19 and 20, 2009, the objective of which was, "to debate how Europe supports and integrates social innovation in its policies, in the wake of its renewed agenda."[59] Forty experts in social innovation took part in the workshop, including Geoff Mulgan and Rushanara Ali, the director and associate director of the Young Foundation. It was also attended by two EU commissioners, the vice president of the European Investment Bank, and the president of the European Commission, José Manuel Barroso, who declared his support for social innovation, thereby demonstrating the importance the EC was attributing to this new line of work. There were similar activities through 2009,[60] and in January 2010 the BEPA officially commissioned the Young Foundation and the Social Innovation eXchange (SIX) global network to draw up a report on social innovation in

59. Bureau of European Policy Advisers (BEPA), European Commission, *Conclusions of the BEPA Workshop on Social Innovation*, 2.

60. Including a June symposium on financing social innovation coordinated by Diogo Vasconcelos from the SIX network. This symposium led, at Mulgan's suggestion, to the "This is European Social Innovation" initiative, the objective of which is to select the ten most important social innovation projects in Europe. This initiative is sponsored by the EC, the SIX network, and the Euclid Network, one of the most important European organizations as regards third-sector studies.

Europe. In contrast to the initial studies of the Young Foundation, which were of a more academic nature (although still influential when it came to British policymaking on the topic), the resulting document, *Study on Social Innovation* (2010), was designed specifically to shape EU policy.[61] And as this study constitutes a key point in the social turn of European innovation policies, I will discuss it briefly.

Young Foundation and Nesta Social Innovation Studies

The Young Foundation has played a very important role in European studies on social innovation, especially following the publication of *Social Innovation* (2007) by Mulgan (with the collaboration of others).[62] The findings of this report were confirmed in subsequent studies, sometimes published in conjunction with Nesta. Another important development was the creation of the SIX network,[63] which gathers information on social innovation from all over the world.

Mulgan's *Social Innovation*, which was originally presented in 2006, mentions numerous historical examples of initiatives that, in his opinion, were significant social innovations: the emergence of labor unions, the cooperative movement (citing the example of the Mondragon Corporation in the Basque Country), kindergartens, sports teams, reading clubs, mutual self-help microcredit unions, philanthropic and charitable organizations, and so on.[64] Mulgan also alludes to different authors who have written on the topic, such as Everett Rogers, Christopher Freeman, Rosabeth Moss Kanter, William Baumol, Eric von Hippel, Bart Nooteboom, Clay Christianson, and John Kao.[65] And he distinguishes several areas in which social innovation processes can be generated today, including social entrepreneurship, design, technology, public policy, cities (urban innovation), social movements, and community

61. See Social Innovation eXchange (SIX) and the Young Foundation, *Study on Social Innovation*.
62. Mulgan et al., *Social Innovation*.
63. See www.socialinnovationexchange.org/.
64. Mulgan et al., *Social Innovation*, 9–10.
65. Ibid., 7. It is worth pointing out that he does not mention Goldenberg, nor indeed any Canadian studies on social innovation.

development.[66] Moreover, the following are mentioned as recent examples of social innovation:

> Self-help health groups and self-build housing; telephone help lines and telethon fundraising; neighbourhood nurseries and neighbourhood wardens; Wikipedia and the Open University; complementary medicine, holistic health and hospices; micro-credit and consumer cooperatives; charity shops and the fair trade movement; zero carbon housing schemes and community wind farms; restorative justice and community courts. All are examples of social innovation – new ideas that work to meet pressing unmet needs and improve peoples' lives.[67]

The last sentence here sums up the general definition of social innovation that Mulgan and his collaborators had used since 2006, although the same work includes a more precise designation of the idea as well: namely, "innovative activities and services that are motivated by the goal of meeting a social need and that are predominantly developed and diffused through organisations whose primary purposes are social."[68] Therefore, the Young Foundation does not just characterize social innovation in terms of its objectives (meeting or resolving social needs or problems) but also requires socially innovative initiatives to have been promoted, at least initially, by agents whose aims are primarily social rather than political, economic, or business-related. Thus Mulgan introduces a more restrictive definition of social innovation than Gold-

66. Ibid., 6.

67. Ibid., 7.

68. Ibid., 8. This definition changes only very slightly in two later publications by the Young Foundation/NESTA, both published in 2007: Mulgan et al., *In and Out of Sync*, and Mulgan, *Ready or Not*. In the latter, Mulgan offers an interesting definition of innovation in the public sector: "The simplest definition is that public sector innovation is about new ideas that work at creating public value. The ideas have to be at least in part new (rather than improvements); they have to be taken up (rather than just being good ideas); and they have to be useful. By this definition, innovation overlaps with, but is different from, creativity and entrepreneurship" (6). This definition matches my pluralist axiological model, according to which in innovation processes the important thing is to create value, although this value can be social, political, economic, cultural, environmental, urban, or of any other kind. Different forms of innovation emerge according to the type of values that are created (or added, in the case of accumulative innovation).

enberg's description, although both views are related.

On the basis of this definition, the Young Foundation identifies three potentiality socially innovative agents: individuals, social movements, and organizations. Specifically, in *Social Innovation*, Mulgan focuses on the third kind of agent, made up of NGOs, voluntary work, charitable institutions, philanthropic foundations, cooperatives, social work, and so on. It is important to point out that Mulgan and his collaborators juxtapose social innovation and business innovation, with the latter being undertaken by companies with a profit motive whose main goal is the acquisition of economic benefit, in contrast to social innovation. Conceptually speaking, with the publication of this document the Young Foundation was configuring an alternative paradigm to the dominant economism model. This was thanks to its close collaboration with Nesta, to the influence its publications had in shaping British innovation policy, and to the creation of the SIX global network that brought together many of the people interested in this social turn in innovation studies. Having said that, the most notable contribution of the Young Foundation to this shift consists of having proposed two definitions of social innovation: one more generic and one more limited. The former was adopted as the starting point by the above-mentioned British government white paper, *Innovation Nation* (2008). The latter is especially interesting for its conceptual and methodological analysis, which I will now pay special attention to.

The first of these definitions is the following, taken from Mulgan's *Social Innovation* (2007): "Social innovation refers to new ideas that work in meeting social goals."[69] *Innovation Nation* paraphrases this and takes from the idea its own definition (previously noted in chapter 2) that can in principle be applied to diverse areas and types of innovation: namely, as "the successful exploitation of new ideas."[70] Insofar as the exploitation of new ideas allows economic, business, social, or any other objectives to be attained, different forms of innovation would accordingly emerge. Given that *Innovation Nation* distinguishes three main sectors of activity, one

69. Mulgan et al., *Social Innovation*, 8.

70. Department for Innovation, Universities and Skills, *Innovation Nation*, 13.

should likewise distinguish just as many forms of innovation: business innovation, that stemming from the public sector, and that which emerges in the "third sector," for example in NGOs and nonprofit organizations. These three sectors interact with one another, but the analytical distinction between each of them is interesting when it comes to drawing up methodologies for the study of different innovation processes in the three sectors. Applying scientific methodology to the study of social innovation implies demarcating precisely the fields in which one seeks to work and beginning with cases that are clearly social in nature, both as regards the agents promoting them and in terms of the objectives of their activities; thereafter, more complex cases, in which achieving political or business objectives are concomitant with social goals, can be considered. This is why I consider innovation stemming from civil society itself as canonical for social innovation studies.

The Young Foundation's narrower definition (again taken from *Social Innovation*) views social innovation in terms of "innovative activities and services that are motivated by the goal of meeting a social need and that are predominantly developed and diffused through organisations whose primary purposes are social."[71] A cursory glance at *Innovation Nation* reveals that the British government white paper follows the conceptual sources of Nesta and the Young Foundation closely. In short, it contends that innovative activity must be directed toward meeting social needs and demands; yet this second definition demands, moreover, that the agents that undertake such innovation should have essentially social objectives in mind. And the three sectors are characterized by their respective values and objectives, whether they are business-oriented, political, or social. This allows one to distinguish social innovation from business innovation, whose ultimate motivation is the maximization of economic benefit.[72] Firms are key agents in business innovation, while governments and public administrations can be innovators in the public sector, which I will address in the next chapter. However, as noted, there is another kind of agent that generates innovation – the third sector – whose

71. Mulgan et al., *Social Innovation*, 8.
72. Ibid.

motives are not political or profit-seeking, but rather strictly social. Analytically, the distinction is clear and it classifies innovation studies according to the agents that innovate and their respective objectives. Even so, the Young Foundation recognizes that social innovation does not just concern civil society:

> Social innovation is not unique to the non-profit sector. It can be driven by politics and government (for example, new models of public health), markets (for example, open source software or organic food), movements (for example, fair trade), and academia (for example, pedagogical models of childcare), as well as by social enterprises (microcredit and magazines for the homeless). Many of the most successful innovators have learned to operate across the boundaries between these sectors and innovation thrives best when there are effective alliances between small organisations and entrepreneurs (the 'bees' who are mobile, fast, and cross-pollinate) and big organisations (the 'trees' with roots, resilience and size) which can grow ideas to scale.[73]

Such strategies, reflected in the bees/trees metaphor, have been very successful, with the trees being represented by public, private, or mixed institutions, corporations, or organizations. To my mind, there is an interconnection between different types of agents, but that does not rule out distinguishing among them and classifying them in different categories. While all three sectors can generate social innovation, and especially if they collaborate with one another, the distinction among them is still important for innovation studies. From an analytical perspective, such a distinction offers the possibility of developing specific methodologies and indicator systems for innovation processes in each of the three sectors, given that the *Oslo Manual* only does so for the business sector. In particular, this distinction allows one to detect new innovative agents, in contrast to traditional models such as Etzkowitz and Leydesdorff's the triple helix model,[74] which only bears in mind the triple university-industry-government focus. Instead of conceiving of civil society as a body that is limited to being a receptor of innovative initiatives, the Young Foundation affirms the existence of innovative agents that come out of civil society itself.

73. Ibid., 4–5.
74. See Etzkowitz and Leydesdorff, eds., "The Dynamics of Innovation."

And this is, in my opinion, one of its main contributions to social innovation studies, independent of the deficiencies in some of its proposals. This is a conception of social innovation that operates "bottom up" as opposed to the traditional models, most of which operate, following the linear model, "top down." There are those who, scarcely without realizing it, hold on to the linear model in social innovation studies and who think that, for example, socially innovative agents are politicians or philanthropists. The pluralist model I propose, in contrast, acknowledges the existence of various forms of social innovation, including that based on the linear model; yet at the same time it also recognizes that this is not either the only or, in general, most interesting form. Starting with this conceptual framework, both the Young Foundation and, above all, Nesta have carried out numerous case studies.

By way of an example, I will cite a joint 2007 report focusing on local innovation systems.[75] This study examines the reasons why some local bodies innovate and tries to explain why some are more innovative than others:

> Traditionally, the voluntary and community sector has been identified as the source of much local social innovation in the UK, especially in the delivery of specialist services for marginalised or vulnerable groups. Conversely, conventional thinking suggests that bureaucracy and a lack of creativity are thought to hamper innovation in local government. However, there is growing interest in innovation in local public service delivery among local authorities and other public agencies working directly with neighbourhoods and communities, as well as those working on more strategic issues like transport.[76]

In order to carry out the study, the authors conducted a survey in various local areas in which it was concluded that 95 percent of the people questioned in positions of responsibility stated that their institutions promoted some type of innovative activity. And the report ultimately concludes that: such initiatives sometimes originate in an administrative body itself, and sometimes as a result of public demand; they are not usually transferred to other local areas; and they do not strictly fall under the technical denom-

75. James and Faizullah, *Making the Most of Local Innovation.*
76. Ibid., 3.

ination of innovation studies, but, rather, they are improvements or developments. Together, both local and social innovation have added value for those that may benefit from them. This is why — although this notion of "placing value" is not of an economic variety but rather meets other kinds of value criteria — one can argue that *different forms of innovation tend to increase the degree to which different values are met*, with the particularity that the solutions offered are new; in other words, they are different to those used previously by those social groups in those social areas.

As I noted already, the Young Foundation accepts three kinds of socially innovative agents: individuals, social movements, and organizations. However, its latest studies have concentrated almost exclusively on organizations, and it has made only brief mention of socially innovative people or social movements as sources of innovation. This is due to a methodological decision that shapes the strategy of its studies and that consists of analyzing above all else social innovation that can be reproduced in other contexts and situations.

Following the trail of the *Oslo Manual* and applying several of its basic concepts to social innovation, the Young Foundation has also addressed social innovation diffusion processes, given that they have no place in markets as such. The diffusion of social innovation is usually local, and sometimes the result of word of mouth, through communities that are receptive to innovative proposals. Voluntary work and cooperation are typical forms of diffusion. If an innovative initiative enjoys considerable social success it can even generate genuine social movements, some of which may go beyond the boundaries of the community from which they emerged. This typically leads to organizational innovations and even to forms of marketing in the media, for example via actions and interventions aimed at publicizing the movement globally, as is the case with Greenpeace. In sum, social innovation also has its own means of diffusion that, generally speaking, do not coincide with those of business innovation whose principal method of diffusion is commercialization.

The Young Foundation does not define social innovation according to the goods and services at stake, but on the basis of the proposals and objectives of the agents that promote it. For there to be social innovation, there must be some kind of agent that seeks to meet some social need or demand and that contributes a new solution to this need or demand. For that reason, the values that guide innovative actions on the part of different agents are decisive, regardless of whether they originate in the public, private, or third sector. When the initial goal is economic benefit, this is business innovation, for all that this innovation may affect social issues and problems. If there is no interest in making any profit and the agent's objectives are primarily social, social innovation can emerge. A company that creates nonprofit foundations can contribute to social innovation through such foundations. The same could be said for public funding of international cooperation and development aid. All of this does not, of course, mean that in whatever the case there may be political or business strategies behind such initiatives. The important thing is that agencies are distinguished from one another on the basis of the values and objectives that guide their actions. A patron of the arts can be an important social and cultural innovator, as can certain international organizations, insofar as social objectives are prioritized over any eventual political or business undercurrents that may be associated with such initiatives. These criteria, although sometimes debatable, allow a register of relevant agents to be established in an innovation system. One of the main problems associated with the study of social innovation is, precisely, that of detecting potentially innovative agents. Any national or regional innovation system has a precise catalog of agents from the primary and secondary sectors. In order to measure innovative activities, it is sufficient to ask these agents if they undertake such activity or not, and if so, what it is. This is how the OECD goes about measuring business innovation. However, there are no such questionnaires for the secondary and tertiary sectors, which limits considerably the reach of innovation studies.

In the case of the tertiary sector, there is usually not even any such register of agents that might be potentially socially innovative. Detecting them is the first task to be carried out, and from then on the survey methodology can be applied. There is another important specific dimension of social innovation that consists of analyzing the results that innovative proposals get. Although the *Oslo Manual* clearly acknowledges that, "many innovations fail,"[77] it does not go on to analyze the success or failure of innovative activities in each firm. In its surveys it only distinguishes three types of innovative activity: successful, ongoing, or abandoned.[78] Obviously, a business innovation process may be abandoned because it has not been successful, but there are also other reasons why such a decision may have been taken: for example, because of a lack of the resources needed to invest in the process, a new strategy on the part of the management, internal problems in the company, and so on.

In sum, besides the intentions, values, and objectives of innovative agents, one must also examine the results they achieve. The ultimate criterion used to assess whether a business innovation is successful or not is the market. In the case of policy innovations, the electorate has the final word. In both cases, there are measurement techniques that allow one to detect the attitudes and evaluations on the part of customers or the public as regards the innovations that originate in the private or public sectors. However, the success or failure of a social innovation depends on the degree of social acceptance it achieves, and this is not typically easy to measure. For this reason, it is worth beginning any study of social innovation with a methodology based on case studies that identifies, in the first instance, potentially innovative agents and that then goes on to analyze good practices and their diffusion. The conclusion here is that social innovation studies should focus on innovative agents and the social sectors that receive the innovation in question, checking the respective evaluations of the innovative proposals and analyzing their diffusion processes. The knowledge transfer that takes place in social innovation processes is demonstrated by the degree of social acceptance for and social appropriation of an innovative proposal. Surveys on public perception and attitudes are indispensable when it comes to measuring social innovation as well as the level of participation in innovative activities (for ex-

77. OECD and Eurostat, *Oslo Manual*, §47, 19.
78. Ibid., §42, 18.

ample, in voluntary work). Besides generation and diffusion, one must also pay attention to this third factor: the social appropriation of innovations.

Another idea to bear in mind consists of the importance one must give to public complaints and discontent. Discontent is one of the driving forces of innovation, as the Young Foundation states clearly when citing the renowned Victorian historian Lord Macauley in one of its reports: "There is constant improvement precisely because there is constant discontent."[79] Social innovations stem from disagreement and protest on numerous occasions, which is why studies of social innovation must always take an interest in any eventual social conflict. As I just remarked as regards failed innovations, innovation studies must pay attention to different negative, as well as positive, factors.

In sum, using the definition given by the Young Foundation as a starting point and the case study methodology established by Nesta, it would appear possible to begin to develop some scientific studies of social innovation. Indeed, the British government has already opted to do so in embracing the same definition and some of these methodological points. For the moment, the dominant methodology consists of studying good practices in social innovation. Introducing measurement methods and a system of indicators for social innovation appears to be more of a long-term task, as we will see at the end of this chapter. It is possible, although not in a comprehensive manner, to distinguish innovative agents according to their objectives and values, as well as to analyze the degree of acceptance social innovation proposals receive by means of case studies. For these reasons, innovation studies can be expanded to include this new and hitherto little-explored field. The traditional dominance of the STEMI model brought with it a bias in innovation studies toward companies and markets. But insofar as there are innovative activities guided by other kinds of values, this has considerably widened the field of innovation studies.

Returning to the contributions of the Young Foundation and Nesta, it is worth paying some attention to the 2010 publication

---

79. Mulgan et al., *Social Innovation*, 9.

Study on Social Innovation.[80] It was commissioned by the EC and published jointly by the SIX network and the Young Foundation.[81] This document marks the growing influence of British agencies not just in the United Kingdom but also in EU innovation policy. I will therefore examine it in some detail because it implies an important nexus between social innovation studies and policies.

Study on Social Innovation offers some significant changes in the positions of Mulgan and his collaborators. Because it was drawn up at the request of the BEPA and the European Commission, it embraces a more flexible conceptual framework and tends toward a more pragmatic synthesis, incorporating other notions of social innovation. The most significant change consists of it recognizing that, as well as the third sector, social innovation can also originate in the private and public sectors, as long as it meets social needs in a new way or confronts significant social problems. This idea was already evident in previous Young Foundation publications, but here it is stated and clarified right from the outset: "It [the report] looks at innovations in all sectors that are achieving high impact and productivity, particularly in relation to goals of better health, education, employment or the environment, rather than focusing exclusively on the public, private or non-profit sectors."[82]

Here, then, we see an interesting methodological and conceptual change. Instead of distinguishing different types of social innovation according to the agents that promote them, which had been customary in previous Young Foundation publications, the 2010 *Study on Social Innovation* is adjusted to be in line with EC objectives and therefore focuses on sectors in which social innovation can have greater social impact and be more productive: specifically, in health, education, employment or the environment. In beginning like this, then, the Young Foundation is evolving in a sim-

80. Social Innovation eXchange (SIX) and the Young Foundation, *Study on Social Innovation*. The people involved in writing up the document were Louise Puldorf and Diogo Vasconcelos for SIX and Julie Caulier-Grice, Lauren Kahn, and Geoff Mulgan for the Young Foundation.
81. See Murray, Caulier-Grice, and Mulgan, *The Open Book of Social Innovation*.
82. Social Innovation eXchange (SIX) and the Young Foundation, *Study on Social Innovation*, 5.

ilar way to Goldenberg and his group in Canada. In other words, it is at least partly recognized that there are three main sources of social innovation: the public, private, and third sectors, but the importance of the environments in which social innovation processes are developed is also underscored in terms of their socioeconomic impact. This being the case, it is stated emphatically that social innovation should be at the center of EU strategy through 2020 as regards innovation policy: "Just as the Lisbon Agenda for jobs and growth focused on innovation, entrepreneurship and the knowledge-based economy, the new strategy for Europe, Europe 2020, must have social innovation at its centre, as a means of stimulating a more dynamic, inclusive and sustainable social market economy."[83]

Does this imply a social turn in European innovation policies? Whatever the case, it is clear that social innovation studies appear to be a real alternative to the dominant paradigm. Social innovation is not even mentioned in the 2000 Lisbon Agenda, which only speaks in terms of technological and business innovation. It was the same story in 2003 when the EC defined its innovation policy. However, by 2010 social innovation was being suggested as a new foundation for European policy.[84] Indeed, what began as the EU's "Europe Flagship 2020 Initiative" was later renamed "Innovation Union" in a 2010 communication[85] in a clear allusion to the British government's 2008 report "Innovation Nation." The EC communication establishes ten courses of action to be taken so that the EU will be highly innovative by 2020, the ninth of which states emphatically: "we must champion social innovation."[86] As a first step in that direction, in February 2010 the EC convened the European Initiative for Social Innovation with the aim of detecting good practices and, in passing, clarifying the notion of social innovation by means of a set of paradigmatic examples.

83. Ibid.
84. I have concentrated here on the EU and its innovation policies, but one could speak equally about a change of focus in conceiving innovation within other bodies such as the OECD, as reflected in the OECD's New Nature of Innovation (2009), which I will not discuss here.
85. See European Commission, *Europe 2020 Flagship Initiative*.
86. Ibid., 3.

In proposing social innovation as an alternative strategic course of action, the Young Foundation and Nesta were being quite prudent and perhaps because of this their proposals were initially successful. Both institutions have, on repeated occasions, criticized the fact that EU innovation policy has been based solely on R&D and have, in turn, made alternative proposals such as focusing on user innovation.[87] Indeed, Nesta published an interesting study in 2010, *Measuring User Innovation in the UK*, which contends that users are an important source of innovation and more efficient, even, than traditional R&D departments.[88] In short, the SIX network and the Young Foundation take a pragmatic line in *Study on Social Innovation*. They state that social innovation is not the same as social enterprise or social entrepreneurship, but also admit that these terms overlap one another.[89]

The strategy of these UK institutions, with which many other European and international groups have been collaborating, through the SIX network, for example, has ended up being a success in Brussels. In May 2014 the BEPA (Bureau of European Policy Advisers) published a report coordinated by Agnès Hubert that incorporated in a large part the proposals of the Young Foundation and Nesta, incorporating them into the design of European policies until 2020.[90] This report affirmed that the economic crisis has demonstrated that technological and entrepreneurial innovation is not enough to resolve major social challenges like employment loss and climate change, which are priorities for the European Commission, which commissioned the report. In order to confront these types of issues, other types of innovations are needed such as social innovations, which are thus defined by Hubert:

> Social innovations are innovations that are social in both their ends and their means. Specifically, we define social innovations

---

87. Social Innovation eXchange (SIX) and the Young Foundation, *Study on Social Innovation*, 13.
88. Flowers et al., *Measuring User Innovation in the UK*. One of the joint authors of this work is Eric von Hippel, who has championed user innovation for twenty-five years.
89. Social Innovation eXchange (SIX) and the Young Foundation, *Study on Social Innovation*, 15.
90. Hubert, *Empowering People, Driving Change.*

> as new ideas (products, services and models) that simultaneously meet social needs (more effectively than alternatives) and create new social relationships or collaborations. They are innovations that are not only good for society but also enhance society's capacity to act."[91]

Dating from the publication of this report in 2010, European innovation studies began to have real influence in European innovation policies, as had happened earlier in the United Kingdom. Even the president of the EC, José Manuel Durão Barroso, elevated social innovation to the main agenda of European policy making, and just at the moment in which the Horizon 2020 program was being drawn up. As a result, in July 2013 the EU programme for Employment and Social Innovation (EaSI) was approved with a budget of 815 million euros for the period from 2014 to 2020. Likewise, other programs related to social innovation were also approved, such as EURES, PROGRESS, and Microfinance facility and Social Entrepreneurship.[92] In total, the EC designed various transversal programs, based in the idea of social innovation, to be developed through 2020. Previously, in February 2013, the EC had published an official document in which it expanded upon and justified the totality of social innovation policies that were to be implemented in the Europe during the period from 2014 to 2020.[93] In this way the events were similar to what happened in 2008 with the UK government: many of the definitions and contributions of the Young Foundation and of Nesta were taken up by the EC with few changed, due to their having been debated and agreed upon in the core of the BEPA, which had organized numerous academic meeting and working groups from 2010 to 2012.

By way of a conclusion to this section, and in a conceptual way, I would point out that the Young Foundation and the Goldenberg and Crises groups in Canada coincide on a key point in regard to sources of social innovation: "Social innovation describes the process of invention, diffusion and adoption of new services or organisational models, whether in the non-profit, public or private sec-

91. Ibid., 7.

92. Seec.europa.eu/social/main.jsp?langId=en&catId=89&newsId=1093 (last accessed December 15, 2013).

93. European Commission, *Guide for Social Innovation*, 2013.

tor."[94] Accepting this criterion implies examining three types of social innovation processes corresponding to each sector. According to the Young Foundation, both the *Stanford Social Innovation Review* (SSIR, a journal and website published by the Stanford Center on Philanthropy and Civil Society, at Stanford University) in the United States and Nesta in the United Kingdom, and even the OECD Forum on Social Innovations, could coincide on this point. This is why I would contend that, in the twenty-first century, social innovation studies appear to have evolved to a point of fairly widespread consensus (which in itself is no small achievement): namely, that social innovation stems, in the first instance, from civil society, but that it can also be generated or implemented by the public sector and by the private sector. Everything depends on the ends and the means. The new definition of social innovations proposed by the SIX network and the Young Foundation in *Study on Social Innovation* I am talking about here is: "Social innovations are innovations that are social both in their ends and in their means."[95] If we add to this the fact that the agents that promote social innovations can come from any of the three already mentioned sectors, the conceptual framework in which the alternative paradigm proposed by British experts from the Young Foundation and Nesta would be sufficiently well defined.

In my opinion, this is an acceptable solution although it still has some drawbacks, especially in its call to the instrumental rationality of means and ends. Philosophically speaking, it would be preferable to speak about values instead of means and ends because both the means selected and the objectives of innovative proposals must be evaluated, whether positively or negatively. In this sense, one should distinguish between economic, social, technological, business, political, legal, administrative, environmental, and other kinds of innovation according to the diverse kinds of values that guide innovative actions and the corresponding processes of diffusion and acceptance. Nevertheless, when it comes to science policies, the means/ends binominal is widely accepted, which is why I think that the consensus definition proposed by the Young

94. Social Innovation eXchange (SIX) and the Young Foundation, *Study on Social Innovation*, 16.
95. Ibid., 17–18.

Foundation is a good starting point from which to advance social innovation studies.

Social Innovation at Stanford University

Although social innovation studies originated in Canada, Great Britain, and Australia, there have been similar developments elsewhere in the world. Indeed, social innovation centers and institutes have been established in numerous countries in recent years. Besides those already mentioned (Crises, the Young Foundation, and Nesta), there are other groups – some of which I have also already discussed – that have made significant contributions to social innovation studies: the Australian Centre for Social Innovation (ACSI) in Adelaide (Australia); InnovationXchange in Melbourne (Australia); the Centre for Social Innovation in Toronto (Canada); the Euclid Network in London (United Kingdom, specializing in the third sector); the Canadian Centre for Social Entrepreneurship (CCSE, now the Canadian Centre for Corporate Social Responsibility, CCCSR) at the University of Alberta (Edmonton, Canada); the Office of Social Innovation and Civic Participation (SICP) at the White House (Washington, DC, United States); the Center for Social Innovation (CSI) at Stanford University (California, United States); the Research Initiative on Social Entrepreneurship (RISE) at Columbia Business School (New York, United States); the Skoll Foundation (Palo Alto, California, United States); the Social Innovation Initiative in Brussels; SITRA, the Finnish Innovation Fund, in Helsinki (Finland); Vinnova, the Swedish Innovation Agency, in Stockholm (Sweden); the Social Development Center (SUS) in Copenhagen (Denmark); the Waag Society in Amsterdam (Holland); the New Zealand Centre for Social Innovation (NZCSI) in Auckland (New Zealand); the Lien Centre for Social Innovation at Singapore Management University (Singapore), the Forum for Social Innovation Sweden at Malmö University (Sweden); the OECD Forum on Social Innovations in Paris (France); the UN-ECLAC Social Innovation Program in Santiago (Chile); the Denok Inn Social Innovation Park in Bilbao (Basque Country, Spain); the ESADE Social Innovation Institute at Ramon Llull University in Barcelona (Catalonia, Spain); Sinnergiak at the University of the Basque Country (Basque Country, Spain); and last but not least, the Centre for Social Innovation (ZSI) in Vienna (Austria). In 2011, the ZSI organized the Challenge Social Innovation 2011 conference, attended by 371 experts in social innovation from 47 countries. In 1991, the ZSI was the first center in the world to incorporate the term "social innovation" in its name.

It would be impossible to report all the conceptual and methodological contributions these institutions have made within the spatial limitations of this work. Moreover, I am sure that the list I have just offered is being added to all the time with new centers, institutes, organizations, networks, and the like. Social innovation is now on the global agenda despite the fact that there are no indicator systems to measure it or, where there are, they are valid for specific areas. For these reasons, in order to conclude this chapter on the emergence of the social innovation paradigm I will limit myself to discussing just one of these centers – the Center for Social Innovation (CSI) at Stanford University in the United States – because its conceptual understanding of social innovation is especially useful for the approach undertaken in this book.

In the spring of 2003, the CSI began to publish the online journal, the *Stanford Social Innovation Review* (SSIR). And in its first volume there is a description of social innovation in the "Editor's Note": "the process of inventing, securing support for, and implementing novel solutions to social needs and problems."[96] Therefore social innovation is defined here according to the existence of social needs and problems. When new solutions are invented to meet such needs and problems and, moreover, when support for and implementation of these is sought, that is when social innovation processes emerge. In this same foundational manifesto, the CSI also points out that social innovation can originate in the public, private, and nonprofit sectors. Revisiting that original document in 2008, James A. Phills Jr., the Claude N. Rosenberg Jr. Director of the CSI, together with Kriss Deiglmeier and Dale T. Miller, the executive director and faculty director, respectively, of the same center, published a very interesting article in which they

96. Quoted in Phills Jr., Deiglmeier, and Miller, "Rediscovering Social Innovation," 36.

proposed distinguishing between social innovation, social entrepreneurship, and social enterprise, a timely task, given the fact that many authors use the three concepts indiscriminately. By means of an example, Phills, Deiglmeier, and Miller contend that the microcredit and microfinance pioneered by the 2006 Nobel Peace Prize winners, Muhammad Yunus and the Grameen Bank in Bangladesh, constitutes a quintessential example of social innovation because it confronts one of the main global problems – poverty – in a new way.[97] They also cite other examples related to the finance sector, displaying a clear interest in social forms of finance, which I will not discuss here.

Phills, Deiglmeier, and Miller state that the concept of social enterprise refers above all to social purpose organizations, while that of social entrepreneurship alludes to the personal qualities of people who start new organizations, with such qualities including "boldness, accountability, resourcefulness, ambition, persistence, and unreasonableness."[98] They also highlight the fact that both terms – social entrepreneurship and social enterprise – have their roots in the nonprofit sector and therefore in principle do not include innovative initiatives originating in the public and for-profit sectors.[99] Third, they offer an important interpretative key – social value – borrowed from a classic article by Greg Dees, which argues that "the underlying objective of virtually everyone in the fields of social entrepreneurship and social enterprise is to create *social value*."[100] And all this leads them to state emphatically, "Ultimately, innovation is what creates social value."[101] Consequently, social innovation studies should, they contend, turn toward the creation of value and, more specifically, social value. However, how does one define the latter?

Phills, Deiglmeier, and Miller discuss up to four different meanings of the term. The first of these is based on the motivations of the agents involved and is that considered by Dees in his

97. Ibid.
98. Ibid., 36–37.
99. Ibid., 37.
100. Ibid. The classic article referred to that influenced the authors is Dees, "The Meaning of 'Social Entrepreneurship.'"
101. Ibid.

pioneering article on social entrepreneurship and social value. The second distinguishes value according to its field of application, concentrating on the social sector, which for many people means NGOs and nonprofit organizations. The third meaning denotes a class of needs and problems, the solution of which concerns a society and therefore generates social value in the event of being achieved. And the final meaning distinguishes social value as something distinct from financial or economic value.[102] Drawing on all this, Phills, Deiglmeier, and Miller define social value as, "the creation of benefits or reductions of costs for society — through efforts to address social needs and problems — in ways that go beyond the private gains and general benefits of market activity."[103]

They thus take it for granted that both public sector activities and those generated by markets are socially beneficial, which could be questioned to some extent as we will see below. Second, they accept that social value refers to meeting needs and resolving social problems, which in principle is true, although this can also be qualified or clarified. And finally, and what to my mind is the most debatable part of their definition of social value, at the same time they distinguish between social value and economic value, in order to define the former they turn to the cost-benefit relationship, the strictly economic roots of which are obvious. For my part, I believe that if one wants to link the notion of innovation to the creation of value, one should develop a theory of value that transcends the sphere of economic value alone and that also takes into account social value as well as other types. This idea will form the core of the philosophy of innovation that I will present in chapter 5.

On the basis of these reflections on social value, the CSI representatives feel qualified to propose two definitions of social innovation. The more complete of these is the following: "A novel solution to a social problem that is more effective, efficient, sustainable, or just than existing solutions and for which the value created accrues primarily to society as a whole rather than private individuals."[104] This is an interesting definition, notwithstanding the debate on the concept of value just discussed, which the au-

102. Ibid., 38–39.
103. Ibid., 39.
104. Ibid.

thors do not explain in sufficiently clear terms. The core of the definition starts from the notion of novelty, logically, but then goes on to choose the solution/problem dichotomy in such a way that social innovations are, in the first instance, new solutions to existing social problems. They then add that those new solutions must be preferable to previously existing solutions, for which they offer four preferential criteria: greater effectiveness, efficiency, sustainability, or justness. For me, it is not sufficiently clarified why they choose these four evaluation criteria and not others when it comes to comparing solutions to social problems. These four criteria are important, but there may be other social values to bear in mind apart from them, such as social peace, freedom, solidarity, or responsibility, to mention four other important social values that could be cited like equality of opportunity. Furthermore, I have my doubts as regards the final requirement, namely that which states that the value created should accrue primarily to society as a whole. Societies are complex entities full of internal tensions and contradictions. Therefore, there are innovations that may benefit specific social groups rather than society as a whole that should not, for that reason alone, be excluded from being considered social innovations. In short, I find the final requirement too restrictive. In any event, I fully accept the connection between the theory of social innovation and the creation of social value, including the suggestion that a social innovation should imply some increase in the level of satisfying certain social values in comparison to previously existing solutions. Novelty on its own is not enough: positive evaluation is required and therefore an improvement. The issue that remains to be debated concerns the axiological criteria that must be taken into account when determining if there has been a rise in or creation of value, but I will address this issue later.

Phills, Deiglmeier, and Miller make another interesting contribution to thinking on social innovation when they distinguish four phases in a social innovation process:

> It is essential to distinguish four distinct elements of innovation: First, the *process* of innovating, or generating a novel product or solution, which involves technical, social and economic factors. Second, the product or invention itself—an outcome that we call *innovation* proper. Third, the *diffusion* or *adoption* of the in-

> novation, through which it comes into broader use. Fourth, the ultimate value created by the innovation. This reasoning gives us the first half of our definition of social innovation: *A novel solution to a social problem that is more effective, efficient, sustainable, or just than existing solutions.*[105]

It would appear as if this four-phase distinction works not just for social innovations but for technological and economic innovations as well. I think it would be interesting to apply it to any form of innovation, including such diverse forms as military, political, cultural, artistic, and environmental innovations, without forgetting epistemic innovations and, for example, scientific discoveries, which also generate value for science. However, perhaps the number of phases needs to be refined somewhat by adding intermediate phases. The language we use (novelties, problems, solutions, evaluation criteria, and so on) should also be adjusted to fit different case studies because it could be the case that the concepts proposed by Phills, Deiglmeier, and Miller are not entirely suitable for some forms of innovation.

Nevertheless, and regardless of the qualifications and objections I raise here toward the conceptual framework proposed by the Stanford CSI, these contributions as a whole constitute another good starting point for studies of social innovation and should be checked against the other studies I have discussed up to this point. By way of an initial reflection, there are at least six important conclusions to be extracted from these ideas associated with the CSI:

1. A novelty is something relative to a previous situation, during which several solutions may have been proposed. For that reason, studies of innovation are always located temporally and must take into account previous states of the art, not just the results that might be obtained.
2. For an innovation to be accepted, whether in markets, public administrations, or civil society, it must have been evaluated positively and to an even greater degree than previous solutions, as pointed out explicitly by Phills, Deiglmeier, and Miller in their above-mentioned definition of social innovation.

---

105. Ibid., 38.

3. Social innovations imply the creation of social value, and other forms of innovation generate or increase other types of value. For that reason, any study of innovation must be based on a theory of values (axiology), which will be presented in chapter 5.
4. Social innovations come from all three sectors – the public, private, and third sector – and especially from collaboration among the three. Various cited Canadian and British authors had already made this point, but the CSI authors clarify it considerably.
5. Social innovations, like innovations in general, are processes that are developed through time, but they can also be thought of as the results that derive from these processes. This dual dimension of innovation must be taken into account, especially because it is not just the end results that can generate value (as the CSI authors state) but also the other phases of the process in question. We should therefore think deeply about the evolving condition of innovation without reducing it to simple results alone, however significant they may be.
6. Provisionally speaking, it is worth maintaining the thesis that there are various types of innovations and not just product innovations. In this regard, the CSI authors' proposals can be implemented by those I previously commented on, including those of the *Oslo Manual*. For that reason, I maintain my initial hypothesis that there are product, process, organizational, and marketing (or communication) social innovations.

These six points can be thought of as the main conclusions of this section, despite the fact that I have also proposed other hypotheses and pointed out certain points of divergence and debate with different authors, including those of the Stanford CSI. I do not claim to have clarified what social innovation is, nor by any means do I fully accept any of the definitions discussed to this point. However, each and every one of them helps to shed some light on the problem in that they all help to explain what social innovation is. By way of conclusion to this chapter, I maintain that social innovation implies or can imply the emergence of a new paradigm for the study of social innovation – and all of this despite the fact that the nascent paradigm (or technoparadigm) has still not been fully constituted. In the final analysis, systematic studies of social innovation only began in the early twenty-first century. And it is to be expected that the emerging paradigm will, with time, be clarified, developed, and consolidated in the years to come.

4
HIDDEN INNOVATIONS

Innovation without R&D

During the twenty-first century, the linear model of innovation has been called into question. The existence of various types of innovation not based on scientific research but on other kinds of knowledge and innovative practices has been repeatedly pointed out. Stephen J. Kline and Nathan Rosenberg had, already in 1986, introduced the "chain-linked model" as a counterpoint to the "science-push model" of Vannevar Bush and his followers.[1] In contending that innovation processes are interactive, disorderly, and subject to changes of many sorts, Kline and Rosenberg anticipated the systemic model developed by Nelson and Lundvall in the 1990s. For his part, in 1988 von Hippel offered empirical data, obtained from various economic sectors, according to which it is not just manufacturers and producers that innovate but also the suppliers, distributors, and users of diverse industrial products.[2] Von Hippel's ideas provoked lively academic debate but did not influence innovation policies in the 1990s, as noted by Patries Boekholt, who contends that European innovation policies during this time were based on four types of instruments: technology transfer mechanisms; instruments to support the establishment of NTBFs (new-technology-based small firms) and university spin-offs; the

1. See Kline and Rosenberg, "An Overview of Innovation."
2. Von Hippel, *The Sources of Innovation*.

setting up of (regional) intermediary structures and programs to help small and medium enterprises (SMEs) to access and absorb technologies; and instruments to secure finance for innovation (seed and venture capital).[3]

The same author likewise points out that Michael E. Porter's ideas about business clusters and his value chain concept were very influential during this time, and indeed are still in use today although we now prefer to speak about *value networks*.[4] Another influential author was Peter Drucker, who picked up on some of Schumpeter's ideas and theorized the knowledge economy, of which innovation is the driving force.[5] In short, the paradigm of technological and business innovation was still fully in force through the 1990s, even though some authors were beginning to notice that scientific research was not the only source of innovation, especially in the case of SMEs. As regards spin-offs, promoting them was important, and it continues to be so today because learning and knowledge transfer are key factors in innovation processes.

Europe was producing quality science, but this research was not generating technology transfer or innovations: a situation the European Commission's *Green Paper on Innovation* termed the "European paradox."[6] In order to confront this problem, many European universities and research centers imitated the American model and created technology transfer offices (TTOs) that, generally speaking, had good results. At the same time, technology parks began to emerge. However, little attention was paid to the knowledge demands of the industrial sector. Knowledge transfer was conceived on the basis of a unidirectional model instead of thinking of it as a systemic necessity among all the stakeholders in innovation processes. In short, the linear model was still in force, although its shortages and drawbacks were being exposed.

---

3. Boekholt, "The Evolution of Innovation Paradigms," 340.
4. Porter's *The Competitive Advantage of Nations* was especially influential.
5. Drucker published two other classic works on innovation: *Innovation and Entrepreneurship* and "The Discipline of Innovation." Here, he underscores the importance of knowledge workers who exist in any company and can be sources of innovation, thereby calling into question the linear model.
6. European Commission, *Green Paper on Innovation*, 5.

Then, in the 1990s and from a very different location – namely, Japan – there emerged a model that clearly went against the science-push model. It was introduced by Ikujiro Nonaka and Hirotaka Takeuchi, who became very influential in the business sector as well as more generally when it came to theories on knowledge management and innovation.[7] Nonaka and Takeuchi contend that a company must manage the implicit knowledge its workforce and management possess, given that externalizing this will generate benefits for the firm. Moreover, knowledge is not confined to R&D laboratories but distributed among a company's entire personnel and mostly in the form of "tacit" knowledge.[8] In order to make sure that such knowledge is transferred throughout a company, Nonaka and Takeuchi propose a SECI (socialization, externalization, combination, and internalization) model,[9] and develop specific knowledge management techniques for internal use in a company. In certain conditions, these four phases of knowledge creation are repeated at higher levels in the company, giving rise to a knowledge spiral that flows from tacit to explicit knowledge.[10] Ultimately, a *creative knowledge company* emerges that does resort to R&D, but also to other sources of internal knowledge and especially to its employees themselves. Some of the Japanese companies that put these management models into practice were highly competitive and innovative, giving rise to the celebrated "Toyota model" that made such an impact on the international market during the late twentieth century. Nonaka and Takeuchi's model was subsequently refined and modified, but it was still one of the first to stand as an alternative to that of generating knowledge and innovation on the basis of R&D alone. In 1997, Linsu Kim published a work that included what would be one of the main models to derive from that of Nonaka and Takeuchi.[11] In this work, taking Korea as an example, Kim demonstrates that a lot of innovation stems from imitation. Contrary to the Schumpeterian notion of innovation based on entrepreneurs with drive, creativity,

7. See Nonaka and Takeuchi, *The Knowledge-Creating Company*.
8. See Polanyi, *The Tacit Dimension*.
9. Nonaka and Takeuchi, *The Knowledge-Creating Company*, 72.
10. Ibid., 71.
11. Kim, *Imitation to Innovation*.

and an ability to take risks, Kim shows that there is another way of innovating that is much more prosaic: namely, that some people just look at what others are doing in other countries, learn how to do what they are doing, and then, in another social and economic context, become innovators because they transfer that learned knowledge to a new economic space. In fact, Kim distinguishes imitation through duplication, which is limited to imitating what others have invented and repeating it in another geographic and economic context; creative imitation, which contributes new dimensions to the original innovation and is not just a copy; and, finally, innovation properly speaking. The first two types are not based on R&D, which is why they offer another example of innovation that does not fit the linear model, despite being very typical.

In sum, at the beginning of the twenty-first century it was clear that there are other forms of innovation within companies and organizations besides R&D. Organizational and marketing innovations are two canonical examples of business innovations not based on R&D, as is creative imitation or simply imitation through duplication. And following publication of the third edition of the *Oslo Manual* (2005), the OECD began to pay more attention to these other forms. With good reason, it defines innovation as something new for firms and not just the world as whole. The innovative might of Japan and Korea is based on this new form of innovation that stems from the R&D that others develop and finance. In more general terms, one might speak in terms of innovation through copying, which comes later and is of a secondary level, but which often contributes greater benefits than the original innovation, precisely because, in copying something original, it is able to generate something new and interesting.

Returning to Europe, one should point out the progressive consolidation of the systemic model, helped in part by its adoption by the OECD, which was especially notable in this organization's *Dynamising National Innovation Systems* (2002). Then in 2005, Sweden established its own innovation agency, Vinnova, to study national, regional, and local innovations systems. Other countries were later in taking such steps, but ultimately national and regional innovation agencies sprung up everywhere. The linear model has often been considered superseded by newer counterparts, but it is still in force in the shape of the underlying notion of the systemic model. Moreover, one must remember that, for some observers, "the innovation system concept is typically a European concept that is not always used so explicitly in non-European countries."[12] Independent of the debate on innovation systems, the evolution of the linear model to the systemic model in Europe has highlighted a series of fields in which hidden innovation – that outside the realm of innovation based on R&D – takes place.

Following the 2006 Aho report, which advocated encouraging a culture of innovation in the EU, R&D is no longer as dominant as it once was. Many other forms of innovation have been detected that are not based on R&D. And the idea that "all forms of innovation need to be promoted" gradually took root within the European Commission.[13] This initial shift in European innovation policy echoes in innovation studies as well. In 2009, for example, the European Agency Innometrics introduced certain indicators to measure innovation not based on R&D as a complement to traditional indicators. While the TPP innovation model is still the EC's preferred model, it is clear that a lot of technological product and process innovations do not stem from R&D but from other sources. And the objective of this chapter is to study some of those.

The Concept of Hidden Innovation According to Nesta

In October 2006 and June 2007, Nesta (the National Endowment for Science, Technology and the Arts), published two reports: *The Innovation Gap* and *Hidden Innovation*. Both documents formally advance the hypothesis that I commented on previously: namely, that "science-based innovation represents only one important dimension of innovation."[14] Indeed, in the same work Nesta estimates that only 6 percent of innovation in the British economy is based directly on scientific research,[15] and that many innovation processes are not being measured by the typical indicator systems (the *Oslo Manual*, the EIS and so on). There thus emerges the hy-

12. Boekholt, "The Evolution of Innovation Paradigms," 345.
13. European Commission, *Putting Knowledge into Practice*, 4.
14. NESTA, *Hidden Innovation*, 4.
15. Ibid.

pothesis of "hidden innovation," understood as a set of innovation processes not detected or measured by the usual indicators. In order to examine this phenomenon, *Hidden Innovation* focuses on six particularly important economic sectors in the United Kingdom: "Oil production, retail banking, construction, legal aid services, education and the rehabilitation of offenders. None of these six sectors invests heavily in formal R&D; nor do they produce many patents. Three represent publicly-funded services that are typically not included in studies of innovation at all."[16]

In demonstrating that very important innovation processes had been developed in these sectors in recent years, and that these processes did not stem from R&D or generate many patents, Nesta opened up a new field of study that since then has expanded. As stated, three of the sectors mentioned are in the public sector. As a result, Nesta calls into question the main dogma of the TPP model: that only companies innovate. Below I will address this new form of innovation – that of the public sector – which is acknowledged as possible by the *Oslo Manual*, but which neither the OECD nor Eurostat have ever studied. For the moment, I prefer to emphasize that, as well as the already mentioned six sectors, in the same document Nesta also detects many other forms of innovation that do not fall under the heading "science-based innovation": one of the most important of these is the culture and creative industry sector, which is very important economically in the UK and has proved itself to be highly innovative. However, it does not fit the R&D-based innovation model nor has it ever been studied as an innovative sector in that way.

As we saw in chapter 3, Nesta defines hidden innovation as "the innovation activities that are not reflected in traditional indicators such as investments in formal R&D or patents awarded. Despite not being measured, hidden innovation often represents the innovation that matters – the innovation that most directly contributes to real practice and performance of a sector."[17] In order to examine this hypothesis, which is extremely important for innovation studies and policies, "UK policymakers have begun

16. Ibid., 5.
17. Ibid., 4.

to address these new and important dimensions of innovation; in particular, the Department of Trade and Industry (DTI) is investigating broader categories of innovation including innovation in services."[18] Nesta draw up a British Innovation Index to detect and measure different types of innovation in the United Kingdom. Logically, this index includes various non-R&D indicators. It has three main components, but the most "economic" of the three — investment in innovation activity — divides the investment made into seven categories: investment in R&D, design, organizational improvement, training and skills development, software development, market research and advertising, and other (copyright development and mineral exploration).[19] And the 2007 *Hidden Innovation* data make it clear that British companies only invested 14.9 percent in R&D, while the respective percentages for the other categories were: 22.1 percent (design), 26.1 percent (organizational improvement), 32.1 percent (training and skills development), 20.2 percent (software development), 14.5 percent (market research and advertising), and 3.5 percent (copyright development and mineral exploration).[20] These empirical data, although they need to be compared with and extended to other countries, are decisive in demonstrating that the linear model of innovation based on R&D is only valid for a relatively few number of firms. Furthermore, this line of inquiry developed by Nesta has given rise to numerous empirical studies that demonstrate the existence of a wide variety of hidden innovations. In short, this is the first major empirical rejection of the linear model that also offers a conceptual alternative — a pluralist model of innovation — as well as a specific methodology to begin to measure other innovation forms.

This plural dimension is also demonstrated in another and, in my opinion, key way for innovation studies: "Innovation policy needs to be sensitive to the sectoral systems of innovation already at work. Where it has a role, innovation policy needs to support systems and focus on promotion and diffusion, not just invention. It is more efficient and effective for government to improve a sector's capacity for innovation than to support the creation of specif-

18. Ibid.
19. NESTA, The Innovation Index, 9.
20. Ibid., 11.

ic innovations."[21] This recommendation by Nesta, if it were to be followed, would alter innovation policy quite radically. Instead of promoting and financing the invention of new artifacts, as R&D policies typically do, it might be more effective to work on other links in the value chain, strengthening other innovation capacities (diffusion, commercialization, organization, links with other companies, developing clusters, and so on). As we will see later, innovation processes generate value, but that can happen at any point in the value network, not just in R&D laboratories (which only form one point in the network, even though their function is very important when it comes to disruptive innovations). However, many innovation processes in companies are not disruptive but accumulative and are based on other forms of knowledge rather than science or technology. The 2008 British government white paper Innovation Nation embraced the Nest hypothesis and, consequently, the Nesta proposals came to shape British science policy and were not just limited to academic debate. In my opinion, this initiated a paradigm shift in innovation studies that has been subsequently developed in recent years and that was clearly reflected in the formulation of the Europe 2020 strategy on the part of the EU, renamed Innovation Union.[22] It is significant that the 2008 British *Innovation Nation* report served as the basis for the European Commission's renaming of this strategy in 2010. And it remains to be seen whether the EC has also adopted the multidimensional focus of innovation. Whatever the case, there are indications that the EU is moving in this direction, at least in certain economic sectors such as the culture and creative industries, which make up another of the major examples of hidden innovation.

Creative Economy and Innovation

The cultural and creative sector (CCS) is the second major example of hidden innovation studied by Nesta. The expression "cultural and creative industries" first emerged in the early 1990s, although not without opposition, and subsequently spread throughout the world. The term implies reducing broader artistic and cultural no-

21. NESTA, *Hidden Innovation*, 6.
22. See European Commission, *Europe 2020 Flagship Initiative*.

tions to industrial civilization, a tendency that has provoked widespread criticism from the cultural and artistic world.[23] In particular, returning to the concept of industry when attempts were being made to promote the knowledge economy – in which the CCS had an important function – seemed paradoxical. Without doubt, important creative, invention, and innovation processes take place in the CCS and these processes are based on knowledge, albeit not on scientific knowledge. Innovation in the CCS must be an object of research given that, generally speaking, it is not based on R&D (although this does not stop it achieving significant social and economic impact). Extending innovation studies and policies in this direction is an important decision and one linked to the shifting paradigm that is taking place, as I contend here. For that reason, it is worth surveying, briefly, the historical debate between the industrial and the cultural, before addressing once more Nesta's proposals and its particular analytical perspective on this sector.

In 1998, Tony Blair's government in the United Kingdom promoted the creative and cultural industries as a technical concept and applied it when obtaining data and statistics in this new sector, the CCS (cultural and creative sector). The DCMS (Department of Culture, Media and Sports) was the body charged with promoting this transformation, which was consolidated in 2001 with a second report on the sector.[24] Australia followed suit and Stuart Cunningham even bettered the United Kingdom's methodology by proposing the model of the creative trident.[25] Cunningham and his colleague in Brisbane, Peter Higgs, applied this model to Australia and obtained important data, which increased by 35 percent

23. In the mid-twentieth century and beyond, Theodor W. Adorno and Max Horkheimer opposed any connection between culture and industry. See Adorno and Horkheimer, "The Culture Industry," and Adorno, "Culture Industry Reconsidered." And in some intellectual circles the term " cultural industry" was thought of as an oxymoron. UNESCO also opposed the use of this expression up to 1980, although later it was the first international institution to create an indicator system by which to measure the culture industry.

24. Department for Culture, Media and Sport (DCMS), *Creative Industries Mapping Document*, 2001.

25. See Cunningham, *What Price a Creative Economy?*

the size of the Australian cultural and creative sector.[26]

At the global level, as the information and knowledge economies have been developed, so the creative industries have grown in strength. In its 2008 report on the creative economy, the United Nations considers the creative industries to be "among the most dynamic emerging sectors in world trade."[27] Specifically, in the period 2000–2005, trade in creative goods and services grew at an average annual rate of 8.7 percent globally and exports of creative services increased by 8.8 percent annually between 1996 and 2005.[28]

According to this report, the creative industries "can be defined as the cycles of creation, production and distribution of goods and services that use creativity and intellectual capital as primary inputs."[29] Therefore, knowledge is thought of in terms of a "primary material" and given a value in that it generates intellectual and artistic goods and services. As a whole, the creative industries are "a vast and heterogeneous field dealing with the interplay of various creative activities ranging from traditional arts and crafts, publishing, music, and visual and performing arts to more technology-intensive and services-oriented groups of activities such as film, television, and radio broadcasting, new media and design."[30] Furthermore, the United Nations argues that "in the most advanced countries, the creative industries are emerging as a strategic choice for reinvigorating economic growth, employment and social cohesion."[31] In a later study concentrating on SMEs in the craft sector, the United Nations distinguishes thirteen subsectors in the CCS: advertising, art and antiques, architecture, crafts, design, designer fashion, film, interactive leisure software, music, performing arts, publishing, television, and radio.[32] The report, therefore, follows the classification of the British government in

26. Higgs and Cunningham, *Australia's Creative Economy*, 5.
27. UNCTAD, *Creative Economy Report 2008*, iv.
28. Ibid.
29. Ibid., 4.
30. Ibid.
31. Ibid., 5.
32. UNIDO, *Creative Industries and Micro & Small Scale Enterprise Development*, 11.

its publications DCMS 1998 and 2001.[33] It also points out, however, that other countries such as Germany and Austria, and the EU itself, have different classificatory schemes.[34] Indeed, in 2000, the EU only distinguished eight such subsectors: books and printing, visual art, architecture, performing arts, audiovisual services, multimedia, cultural heritage, and archives and libraries[35]. Since then, there has been widespread debate within the EU with the aim of establishing a common conceptual framework and making the respective statistical data compatible both in the cultural and the creative sector.

Meanwhile, economic growth in the cultural sector has been particularly strong in Europe. The KEA study *The Economy of Culture in Europe* (2006) offers significant data that continue to be a reference point. The same study also proposes a new model by which to analyze the CCS: the "radiation process."[36] This model has been adopted by many researchers on the topic because it allows them to structure the data and classifications of different countries in a similar way while maintaining the specificity of each. The model incorporates a core made up of the CCS or "arts field" (the visual arts, performing arts, and heritage), followed by two outer circles around this core consisting of "cultural industries" (film and video, television and radio, video games, music, and books and press) and "creative industries and activities" (design, architecture, and advertising). As regards the core arts field, the visual arts include painting, sculpture, and photography, but also crafts, which is quite surprising. Thereafter, the performing arts include theater, dance, the circus, and festivals; heritage includes museums, libraries, archives, and archeological sites.[37] Therefore, the KEA classification incorporates all the subsectors proposed by studies published in the UK and Australia but differentiates between the cultural and

33. Department for Culture, Media and Sports (DCMS), *Creative Industries Mapping Document*, 1998 and 2001.

34. UNIDO, *Creative Industries and Micro & Small Scale Enterprise Development*, 16–17. After British studies (DCMS 1998 and 2001), Austria was the second European country to examine its own CCS in 2003. See KMU Forschung Austria (KMFA), "Creative Industries in Austria."

35. See KEA European Affairs, *The Economy of Culture in Europe*, 51.

36. Ibid., 53.

37. Ibid., 56.

creative industries, which is acceptable to other European notions about the sector, such as those in France and Germany. Moreover, not everything is industry in the cultural sector, which is why a cultural core is identified that is not reduced to industrial value. I believe this concentric model may be acceptable for innovation studies in the CCS, although with three important qualifications:

First, the thirteen subsectors in both the British and United Nations classifications do not completely cover the CCS. To my mind, the aforementioned core should also include languages. It is also worth trying to judge their impact and quantify them economically, even if this may be a difficult task.[38] Second, cultural and creative activities generate added value in other economic sectors, including scientific and technological research. Indeed, in a knowledge economy one should always analyze the ACT (art, science, technology) sector. The same could be said for the influence of the core cultural sector (museums, art, heritage, and so on) on tourism, as well as other cultural and creative industries on entertainment. Put another way, as regards innovation processes, the CCS is transversal and favors such processes in different social and economic sectors. For that reason, it is not enough to just quantify its direct impact in the economy of a country of a region; one should also estimate its hidden impact. Third, the concepts of "sector" and "subsector" that are typically used to analyze culture and creativity can be questioned for various reasons, including for example on the basis of the methodology employed. In other words, a creative or cultural activity may often have an impact in not just one but several subsectors.

Having offered these qualifications, let me now move on to the empirical data offered by the KEA study, *The Economy of Culture in Europe*. In 2003, the CCS contributed 2.6 percent of EU GDP. However, to this one should add the particular feature that the sector grew by 19.7 percent in the period 1999–2003, with this figure being 12.3 percent higher than the growth of the European economy in general during the same period. Moreover, as regards

38. In the case of Spanish, an initial attempt is made by Martín Municio, ed., *El valor económico de la lengua española*. This study contends that, in 1997, the Spanish language accounted for 14.64 percent of Spain's GDP and it estimates that this would increase to 15 percent by 2011.

employment in the cultural sector, it grew 1.85 percent in the period 2002–2004 even though employment in the EU as a whole during the same period fell. In sum, in 2004 almost six million people worked in the cultural industries or 3.1 percent of the total employed population in the twenty-five countries of the EU.[39] We can conclude, therefore, that the cultural sector is one of the most dynamic sectors in the EU information and knowledge economy. What is more, it generates a lot of employment.

Spain is a country in which growth in cultural employment was most pronounced in the period 2001–2006. Of the twenty-five EU regions with the highest average annual employment growth in the CCS, ten were in Spain, with percentages ranging from 5 to 8.5 percent annual growth in employment. And the same could be said for the economic growth of the sector as a whole, not just in terms of employment. In the region of Asturias, to cite a prominent example, the CCS growth rate was 8.42 percent for the period 2001–2006 while the regional economy as a whole grew by only 3.5 percent in the same period. In the Basque Country, meanwhile, the CCS growth rate was 6.51 percent for this same period as opposed to 2.52 percent for the regional economy as a whole.[40]

These data indicate an important tendency in knowledge-based economies. In Europe, as a whole the cultural and creative industries are more important economically than agriculture, livestock farming, and fishing put together. They also surpass the textile industry and are almost at the same level of the chemical industries. The cultural and creative industries make up an emerging sector that is worth paying some attention to because they offer the potential for growth in the transition from industrial societies to information and knowledge economies.

As regards the behavior of this sector in Europe in recent years, unfortunately there is no up-to-date data available for the EU as a whole. However, some private agencies have carried out estimates. In March 2010, for example, one such report was published that estimates that "in 2008 the European Union's creative industries, based on the more accurate and comprehensive definition, contrib-

39. KEA European Affairs, *The Economy of Culture in Europe*, 6.
40. Power and Nielsén, *Priority Sector Report*, 11.

uted 6.9%, or approximately €860 billion, to total European GDP, and represented 6.5% of the total workforce, or approximately 14 million workers."[41] If this were the case, the relative strength of the sector would be quite considerable and it would double the KEA estimates for 2004. For that reason, one should approach these data with caution, especially seeing as though they are not broken down or justified in any detail (which the KEA study does do). Even so, there is no doubt that the CCS is very important economically, whether directly or indirectly.

The growing importance of the CCS in industrialized countries has raised the need, as was the case with R&D and innovation, to *measure culture and creativity* with the aim of guiding public policies and business strategies in the sector. Measuring culture and creativity is not easy due to their complexity and diversity, just as it is likewise difficult to measure innovation. As noted, UNESCO was the first organization to take up the challenge and establish a conceptual and methodological framework for developing cultural statistics at an international level.[42] In 1980, it organized an International Conference on Communication in Acapulco, Mexico, at which the issue was addressed for the first time. Two years later, a symposium was held in Austria under the title "Framework for Cultural Statistics" (FCS), the end results of which were presented by UNESCO in 1986. From that moment on, there was a common frame from which to draw up cultural statistics. This frame was adopted by most countries when it came to their own national statistics, although with their own variations and specific qualities, hardly surprising, given the different dimensions of culture throughout the regions of the world. However, UNESCO modified its measurement procedures regarding culture in 2009 in order to adapt them to the huge impact that ICTs had had on different cultural practices. The other significant factor that obliged

41. TERA Consultants, *Building a Digital Economy*, 5.

42. From an academic perspective, by the mid-1980s economists had generally come to accept the sub-discipline of cultural economics, with the designation Z10 being given by the *Journal of Economic Literature*. The new group of economists in this field subsequently came to be associated with the *Journal of Cultural Economics* and the Association for Cultural Economics International (ACEI).

it to revise its framework of cultural statistics was globalization, which was not just economic but also cultural. Finally, in recent years important subsectors also emerged, such as design, fashion, and video games, which did not fit the original statistical framework and which in some countries were very important economically, therefore pointing to the need for new measurement criteria. In short, UNESCO's recent modification of its conceptual, classificatory, and methodological framework has given rise to significant changes in the structure of cultural data.[43]

The basic conceptual framework, then, is the following: In order to construct cultural statistics, UNESCO distinguishes different "cultural domains" as well as several phases in a cultural value chain, or what it defines as a "culture cycle." Its measurement instruments are organized on the basis of a matrix that combines eight cultural domains and five phases in the culture cycle. In order to draw up cultural statistics, then, one needs specific statistics and indicators for each of the forty possible compartments contained in this matrix.[44] A cultural map of a country can thus be drawn up on the basis of this matrix and the indicators contained therein, and then compared to similar maps for other countries. In addition to cultural domains, UNESCO takes into account "related domains," which "consist of other economic and social activities that may be considered 'partially cultural' or that are more often regarded as being 'recreational or leisure' rather than 'purely cultural.'"[45]

The latter clarification is important because one of the particular features of the CCS is that it, besides generating value in itself as a distinct sector, also favors the creation of value in other economic, cultural, and social sectors. Cultural and creative statistics must always take into account related domains and not just the actual cultural domains. Furthermore, UNESCO distinguishes five stages in any culture cycle:[46]

1. Creation: the originating and authoring of ideas and content, which comes from artists, sculptors, writers, compos-

43. See UNESCO, *2009 UNESCO Framework for Cultural Statistics.*
44. Ibid., 85.
45. Ibid., 10.
46. Ibid., 19–20.

ers, designers, and so on, including the making of original work.

2. Production: the reproduction of original work with the appropriate tools to do so (musical instruments, for example), including the infrastructure (factories, printing presses, and so forth) and processes used in their manufacture (the printing of newspapers or books, for example).
3. Dissemination: the delivery of reproduced cultural products to consumers through different channels of distribution (the retail and rental of recorded music or film distribution, for example).
4. Exhibition/Reception/Transmission: the places of consumption or provision of live cultural experiences (theaters, museums, galleries, libraries, radio, television, and so on).
5. Consumption/Participation: the use of cultural goods and services by people and participation in cultural activities and experiences (book reading, dancing, listening to radio, and so forth).

In distinguishing these five stages, UNESCO makes an important contribution to defining the basic structure of the value chain in the CCS. Its model was later refined by the addition of domains and stages in the cycle. As regards domains, initially six cultural domains were identified (cultural and national heritage, performance and celebration, the visual arts and crafts, books and press, audiovisual and interactive media, and design and creative services), together with two related domains (tourism and sports and recreation).[47] As regards the cultural sector, it was gradually added to from the creative sector, and the number of domains increased, as did the stages in the culture cycle, sometimes.

This hybridization between the cultural sector and creative industries was not a UNESCO initiative but instead originated in the United Kingdom In the mid-1990s the British government Department for Culture, Media and Sport (DCMS) created a group of experts known as the Creative Industries Taskforce. In 1998, it presented a report, the *Creative Industries Mapping Document*,

47. Ibid., 23–24.

which proposes and develops a new methodology for measuring the CCS jointly.[48] A second edition of the report was then published in 2001, in which the concept of creative industries is defined explicitly on the basis of intellectual property rights, implying a radical shift in relation to the criteria UNESCO used to measure culture. According to this latter document, the creative industries are those "which have their origin in individual creativity, skill and talent, and as having the potential for wealth creation through the generation and exploitation of intellectual property."[49] Moreover, they are classified by the DCMS taskforce into the thirteen subsectors or domains listed above: advertising, architecture, the art and antiques market, crafts, design, designer fashion, film, interactive leisure software, music, the performing arts, publishing, software and computer services, and television and radio. With the publication of this document, though, the previously established UNESCO criteria began to be questioned in several English-speaking countries. For example, terms like the "creative economy"[50] and the "creative industries" were increasingly employed rather than just "culture" or the "cultural industries." And other countries began to follow the British example, while at the same time introducing other specific variants in their own statistics, such as New Zealand (2002), Australia (2003), Singapore (2003), and Hong Kong (2003). Meanwhile, a study of the creative industry sector based on the British model was also published in Austria in 2003.[51]

There was an important milestone in 2007 with the publication in Australia of a study on the creative economy by Peter Higgs and Stuart Cunningham that introduced another major example of hidden innovation. These authors demonstrate empirically that cultural and creative activities generate value in other economic sectors. And, indeed, they contend that scientific and technological research can also be understood as a creative activity, which has given rise in recent years to an interesting convergence between

48. See Department for Culture, Media and Sport, *Creative Industries Mapping Document* (1998).
49. Department for Culture, Media and Sport, *Creative Industries Mapping Document* (2001), 3.
50. See Howkins, *The Creative Economy.*
51. KMU Forschung Austria (KMFA), "Creative Industries in Austria."

the arts, sciences, and technology (the AST system), especially after the introduction of ICTs into technoscientific activity.[52] In any event, the multiplying effect of creative activities when it comes to generating value and innovation in other economic sectors is undeniable. This is why the European Commission made a series of important decisions regarding culture from 2006 on that resulted in official approval, on November 23, 2011, for the Creative Europe program (to take place between 2014 and 2020). These decisions included EU approval for the Agenda for Culture in 2007, within the framework of which there was encouragement for initiatives such as creating the European Creative Industries Alliance (ECIA) in February 2010. The ECIA was charged with developing a European agenda for culture and creativity, including a remodeling of cultural statistics. And in March 2010, the European Cluster Observatory (ECO), based in Stockholm, Sweden, published an extensive report on the relative strength of the CCS in different European regions and cities. Among other important ideas, its executive summary states that:

- In 2006 European creative and cultural industries firms employed a total of 6.5 million persons.
- Regions with high concentrations of creative and cultural industries have Europe's highest prosperity levels.
- Large urban areas and capital city regions dominate the creative and cultural industries, but some city regions do better than others. The super clusters London and Paris stand out, followed by Milan and Amsterdam.[53]

For its part, the European Commission published in 2010 its *Green Paper: Unlocking the Potential of Cultural and Creative Industries*, which served as the basis for drawing up the aforementioned Creative Europe program (2014–2010). The green paper provides definitions for the cultural and creative industries, attempting to

52. Some authors describe the ACT subsystem as one of the most important dimensions of the knowledge economy. See, for example, the report by the Fundación Española para la Ciencia y la Tecnología (Spanish Foundation for Science and Technology), *Libro Blanco sobre Arte, Ciencia y Tecnología* (white paper on art, science, and technology).
53. See Power and Nielsén, *Priority Sector Report*, 2.

resolve the debate among authors based in the United Kingdom, Germany, France, and other European countries. Having been adopted through consensus, the following may be considered as the current "official" EU definition of the cultural industries:

> "Cultural industries" are those industries producing and distributing goods or services which at the time they are developed are considered to have a specific attribute, use or purpose which embodies or conveys cultural expressions, irrespective of the commercial value they may have. Besides the traditional arts sectors (performing arts, visual arts, cultural heritage – including the public sector), they include film, DVD and video, television and radio, video games, new media, music, books and press.[54]

And as regards the creative industries, meanwhile, the definition is as follows:

> "Creative industries" are those industries which use culture as an input and have a cultural dimension, although their outputs are mainly functional. They include architecture and design, which integrate creative elements into wider processes, as well as subsectors such as graphic design, fashion design or advertising.
>
> At a more peripheral level, many other industries rely on content production for their own development and are therefore to some extent interdependent with CCIs. They include among others tourism and the new technologies sector. These industries are not explicitly covered by the concept of CCIs used in this Green Paper though.[55]

As one can see from this, the EC avoids the more profound debate on the concept of the term "creative industry," which as noted above the British government DCMS bases on the notion of intellectual property along with the World Intellectual Property Organization (WIPO). Indeed, it was clear from the EC Measuring Creativity conference in Brussels in late May 2009 that there were major discrepancies among countries when it came to putting the cultural and creative sectors together as one or not. For these reasons, the EC green paper opts to respect UNESCO's conceptual framework, although updating it somewhat. Simultaneously, a group of experts from Eurostat was commissioned to reshape the

54. European Commission, *Green Paper*, 5.
55. Ibid., 6.

statistical framework of culture and creativity in Europe. This group, named ESSnet-Culture, worked on the project between 2010 and 2011, when it published a long draft report.[56] Here, the group proposes organizing European statistics in the CCS along the lines of ten domains and six functions or stages. The ten domains are cultural heritage, archives, libraries, books and press, visual arts, performing arts, audiovisual and multimedia, architecture, advertising, and art craft. And the six functions or stages are creation, production/publishing, dissemination/trade, preservation, education, and management/regulation.[57]

Therefore, the Eurostat experts' report does not resolve the more profound debate, and we will have to wait for a policy decision to be made that might determine exactly how many cultural domains are to be accepted within the CCS and which stages actually define the culture cycle. Whatever the case, the forty initial compartments in the UNESCO matrix (see above) have now been extended to sixty by the Eurostat 2011 draft report. This gives some idea of the profound changes that are taking place when it comes to measuring culture and creativity in Europe. As regards the indicators that are widely accepted as standard in Europe, the ESSnet-Culture group acknowledges eleven "key" indicators that offer data on the direct impact of culture in the economy and seventeen "indirect" indicators aimed at analyzing indirect participation or "secondary effects" (or "spillover effects") in other economic sectors.

The way I see it, this is key to understanding EU interest in cultural and creative activities, as they come to be termed in the Eurostat 2011 draft report: namely, the notion that they do not just generate wealth in the CCS – regardless of whatever subsectors or domains might be distinguished there – but that they also favor the creation of value in other economic and social sectors. This is why their relative significance within the GDP of countries is high and increasing, insofar as creativity is one of the driving forces of knowledge economies and innovation.

Every study and report coincides in its approval for the Cre-

56. ESSnet, *Project ESSnet Culture.*
57. Ibid., 31–32.

ative Europe program (2014–2010), passed by the EC in 2012 and now underway. This program forms part of a new wider European innovation policy — "Innovation Union" — that fully recognizes the importance of the CCS for innovation, as well as its growing economic strength. According to the report presented by Ann Branch, head of unit for the culture program and actions at the European Commission, the CCS contributes 4.5 percent of the EU GDP and 3.8 percent of the EU workforce.[58] Yet its indirect strength is even greater due to the impact of the cultural and creative industries on innovation processes that take place in other economic sectors: "Beyond this direct contribution to jobs and growth, these sectors trigger spill-overs in other areas such as tourism, content for ICT and provide benefits for education, social inclusion and social innovation."[59] As noted, this is the main argument of the EU when it comes to underscoring the importance of the CCS for innovation studies and policies. The CCS is very different to R&D and depends administratively on other departments and ministries, and it generates value in other sectors, including R&D, which is also a creative activity. For that reason, one might conclude that *the CCS is transversal in any innovation system*. This is where its specific importance comes from but also explains why it is difficult to measure. In any event, ignoring creative activities when it comes to drawing up innovation policies is a strategic error that certain countries have attempted to correct in recent years. The most obvious example of this is the United Kingdom, but specific mention should also be made of Hong Kong, Australia, Canada, and Sweden.[60]

There is not the space here to deal with all these countries and reports, even though they are of interest. For that reason, I will once again focus on Nesta, albeit briefly, and its studies and reports in relation to the CCS. In one such publication, the policy briefing *Beyond the Creative Industries* (2008), it is argued, "The UK has the largest creative sector in the European Union, and probably the largest in the world relative to GDP. The Department for Culture, Media and Sport (DCMS) estimates that in 2005 the creative in-

58. European Commission, *Creative Europe*, 2.
59. Ibid.
60. On Hong Kong, see Hui, *Baseline Study of Hong Kong's Creative Industries*.

dustries accounted for 7.3 per cent of total UK Gross Value Added (GVA), and grew at 6 per cent per year on average between the years 1997 and 2005."[61] If such data are correct, then this is one of the most important forms of hidden innovation. It would appear, then, that the CCS is well placed to become one of the most highly innovative sectors in the economy.

In 2008 as well, Nesta published a more extensive study, *Hidden Innovation in the Creative Industries*, authored by Ian Miles and Lawrence Green. It focuses on four specific subsectors – video games, product design, advertising, and independent broadcast production – as well as a new form of hidden innovation based on creating new experiences: "Innovation often occurs when those producing creative content respond to the experiences of consumers and users, and make changes to their offer as a result."[62] Other forms of (already known) hidden innovation are also analyzed in the study, such as organizational innovation in the CCS as a result of user participation in designing products and processes and innovation based on new combinations of technologies and processes. Above all, the authors recommend that innovation studies and policies deal closely with the CCS due to it being one of the most characteristic sectors among the different forms of hidden innovation. Indeed, for them, "innovation and creative industries studies have rarely been brought together in a systematic way."[63] Finally, they suggest four categories of hidden innovation: innovation that is the same or similar to activities that are measured by traditional indicators, but which is excluded from measurement; innovation without a major scientific/technological basis, such as innovation in organizational forms or business models; innovation created from the novel combination of existing technologies and processes; and locally developed, small-scale innovations that take place "under the radar" and are therefore unrecognized or accounted for.[64]

These considerations demonstrate that the issue of hidden innovation is quite extensive, complex, and protean. The CCS is

61. NESTA, *Beyond the Creative Industries*, 1.
62. Miles and Green, *Hidden Innovation in the Creative Industries*, 6.
63. Ibid., 14.
64. Ibid., 16.

a clear example of this because most of the innovation processes produced in that sector cannot be detected by the dominant methods for measuring innovation such as the EIS or the *Oslo Manual*, which are traditionally inclined toward technological, manufacturing, and R&D-based innovation. The emergence of a new paradigm in innovation studies is clearly demonstrated in the CCS. In principle, no one would question the idea that innovation and creativity are connected, which is why the creative industries and the creative sector must be of interest to innovation studies and policies. However, it is not at all easy to measure creativity, as I will demonstrate below. There, I will offer some more extensive arguments to explain why there is debate in Europe about culture, creativity, and innovation.

Creativity and Artistic Innovation

In this section I will focus on artistic practices, some of which are highly innovative. The arts have a key function in the debate over creativity and innovation, precisely because they are atypical in relation to R&D activities. Obviously they generate innovations, including the art market and creative industries. However, artistic research is distinct from scientific research, especially as it does not turn to the peer system review of academia to evaluate the quality of artistic work. For all that the arts and sciences find mutual benefit in maintaining close relations — and recognizing the fact that nowadays there is hybridization between the arts, sciences, and technology — innovation processes in artistic environments are very different to those that take place in R&D laboratories. The arts imply a clear anomaly for the second dogma associated with the linear model. In other words, literature, painting, sculpture, architecture, music, dance, theater, and film have their own systems of knowledge and innovation transfer that are not always equivalent to the TPP model of the *Oslo Manual*. Current indicator systems have paid little attention to artistic innovation (in the plastic arts, music, design, fashion, and so on) and this despite their undoubted economic importance. One only need think about the record industry and its dizzying innovative rhythm (the development of hit singles, the charts, downloadable music, and so on).

Artistic performance series and literary works such as the Harry Potter series by J. K. Rowling are also worth considering in innovation studies in what is referred to as the *economy of experience*. In short, artistic innovation networks only partly resemble the dominant technoscientific model. Indicators of success and social acceptance for an innovation are sometimes economic in nature (for example, a sales index) but on other occasions they have more of a social dimension: for example, concert or exhibition attendance rates, media influence, international impact, and so on. In any event, procedures to evaluate the quality of artistic innovation are different to those of science, just as they for their respective evaluation criteria and corresponding value chains.

Let us pause for a moment and focus on film, with the aim of analyzing cinematographic innovation processes in greater detail. It goes without saying that the "seventh art"[65] has generated a powerful industry in which a good script is usually decisive in the competitive struggle for success among many different productions. Screenwriters, directors, and actors – and not just the producers who invest money in movies – are significant agents in cinematographic innovation networks. Other times, the success of a movie stems above all from its special effects, set design, or costume design, to mention just three technologies that contribute to the accomplishments of film productions. The Oscars, awarded by the Academy of Motion Picture Arts and Sciences (AMPAS), as well as numerous similar awards at international film festivals, demonstrate that in this case, too, there are procedures for evaluating the quality of both movies and the actors who appear in them. Such evaluation processes are different to those of science, but they do exist. Nevertheless, the crucial point here is social acceptance when the public watches films, either in movie theaters, on video, or through the Internet. Box-office receipts, the duration of films in movie theaters, video/DVD rentals and sales, and online downloads are other kinds of cinematographic innovation indicators. The task of constructing innovator systems for innovation processes in artistic fields, when attempted, should bear in

65. The term "seventh art" was applied to cinema by one of the first film theoreticians, the Italian Ricciotto Canudo.

mind these kinds of measurement procedures. Unfortunately, the *Oslo Manual* does not consider such issues. One can only conclude, then, that today this document only measures a portion of all innovation, and moreover through very restrictive criteria of an exclusively economic- and business-related nature.

Moving from the fine arts and film to other innovative activities not based on scientific research, the previous argument becomes even more valid. Take fashion, for example, or the design arts in general, which generate both important markets and significant international companies. One could also cite gastronomy, which is at times highly innovative and which has a major and direct impact in society as a whole, without forgetting the sheer volume of business it generates. And there are also innovative activities in the "lesser arts," sometimes with an international dimension and other times at a regional or local level. Analyzing such innovation systems from a dual perspective – both social and economic – is essential.

In conclusion, cultural and artistic changes and developments are also a source of innovation, which cannot be reduced to technoscience alone. Both knowledge and development allow for more than just one meaning or form. It seems obvious to say so, but the truth is that most theoreticians of innovation have only ever examined those innovations that, stemming from scientific research and taking shape in technological prototypes, are channeled by companies toward society through markets. In continuing in this way, then, their models are a fortiori insufficient.

Having said all that, there remains the thorny issue of how to measure creativity and innovation in the diverse artistic domains. As regards this problem, Richard L. Florida has been especially influential in the early twenty-first century, with his ideas having helped define the European Creativity Index. To finish up these reflections on innovation in the CCS, I will offer some brief critical observations of Florida's proposals.[66]

These proposals are extremely ambitious given that Florida suggests, in *The Rise of the Creative Class, and How It's Transforming*

66. In particular, I address two of Florida's works: *The Rise of the Creative Class* and *Who's Your City?*

Work, Leisure, Community and Everyday Life (2002, and additionally in the preface to the), that we have entered into the "creative age";[67] that creativity is what distinguishes humans from other species;[68] and, last but not least, that "because creativity is the driving force of economic growth, in terms of influence the Creative Class has become the dominant class."[69] Therefore Florida is audacious enough to propose an alternative economic theory to those of the great economists in history, including Schumpeter. Let us recall that for Schumpeter, for example, innovation was the driving force of economic growth, as we saw in chapter 1.

In order to justify such bold proposals, Florida offers spectacular data on the United States, although he does not clarify exactly what methodology was used to calculate such figures: "The wealth generated by the creative sector is astounding: it accounts for nearly half [of] all wage and salary income in the United States, $1.7 trillion dollars, as much as the manufacturing and service sectors combined."[70] He goes on to conclude:

> Traditional models that say that economic growth comes from companies, or jobs, or technology are thus incomplete. I agree wholeheartedly that technology is important. But other factors come into play. Technology – measured by innovation and high-tech industry concentration – figures into my model as one of the "3 Ts" needed for growth. Talent is the second T – not "human capital" as usually measured (by numbers of people holding higher education credentials) but creative capital, which is talent measured functionally, by the numbers of people actually in creative occupations. The third T is tolerance. Places that are open and tolerant have an edge in attracting different kinds of people and generating new ideas.[71]

The 3T model is clearly elitist because it establishes a boundary between those who are talented and those who are not. Because the former are the driving force of economic growth and wealth, the latter appear to be condemned to be subordinate to them.

67. Florida, *The Rise of the Creative Class*, 10.
68. Ibid., 4.
69. Ibid., xxvii.
70. Ibid., xiv.
71. Ibid.

However, it is not at all clear what talent actually is. Even though it is debatable, the notion of human capital can be measured, given that it depends (among other factors) on the number of graduates and doctors in a particular society. However, it appears that Florida bestows the quality of talent on people who have secured "creative occupations," thus implying that, unlike other economic sectors, there are no inept people employed in the creative sector! Moreover, in order to delimit exactly what a creative occupation is, one would need to define precisely what the creative economy and creative industries are – a point on which we are a long way from reaching any consensus. In short, as an indicator for measuring purposes, the T of talent is a very vague and confusing criterion.

Higgs and Cunningham's creative trident model makes a careful distinction between creative and auxiliary activities by which the latter, although also developed in the creative sector, cannot strictly be considered creative. Thus, arguing that a person is creative merely because he or she is employed in the creative sector is an error on Florida's part. Both Schumpeter (who Florida mentions on more than one occasion) and Drucker make a big effort to distinguish innovative entrepreneurs from those who are not, making it clear that ultimately there are very few true innovators. Believing that everyone who works in science and engineering, architecture and design, education, the arts, music, and entertainment is creative per se is equivalent to thinking that the creative professions imprint people with character and that they select, by some mysterious criteria, people who are authentically creative, leaving non-creative people without any work in any of these sectors.

In sum, the methodology underlying Florida's arguments is open to criticism: he selects a number of economic activities in the standard occupational classifications (SOC) system (the US government system for classifying occupations, used by the US census for example), counts the number of people who carry out these activities, adds them up, and uses this total to calculate the number of people who make up the "creative class." And, for whatever reason, the "super-creative core" of this same class includes people who work in computers and mathematics; different forms of engineering; the social, physical, and life sciences; education, teaching, and reading; and the arts, design, entertainment, sports, and the media. All this is stated without giving any reasons or bases as to why some of these professions are included in one category, others in another, and others still not included at all.

One could say the same about his later book, *Who's Your City? How the Creative Economy Is Making Where You Live the Most Important Decision of Your Life* (2008). According Florida, nation-states are no longer the driving forces of economic growth and have been replaced by creative cities. He borrows the notion of "city-region" from Jane Jacobs and this is undoubtedly very interesting.[72] What remains questionable is the 3T criterion previously used by Florida to characterize creativity, given the extremely vague and ambiguous nature of the indicators suggested to measure the three "T"s in question (technology, talent, and tolerance). Constructing a system of indicators to measure creativity, or where appropriate innovation, is undoubtedly a creative task, but it is much tougher and rigorous than Florida thinks. Ultimately, I think his main conceptual error consists of having focused on invention (which he terms creativity) and forgetting the Schumpeterian criterion by which one must distinguish between invention and innovation. In *The Rise of the Creative Class*, Florida states explicitly, "In my formulation, 'knowledge' and 'information' are the tools and materials of creativity. 'Innovation,' whether in the form of a new technological artifact or a new business model or method, is its product."[73] I consider this to be one of the key elements in his thinking. In contrast to Schumpeter, who was a precise and rigorous thinker (without ignoring the fact that he may have been wrong in some of his arguments), Florida only focuses on the initial phases of innovation processes, in which invention and creativity dominate. Yet as almost all authorities on innovation argue, most innovative ideas fail and only very few of them are truly successful. This is the first problem with creativity, especially when it is confused with imagination. One might be very creative, imaginative, and inventive, but in order to generate an innovation one needs other abilities

72. *Florida, Whi's Your City?*, 43–44. Florida specifically takes the notion of the "city-region" from Jacobs, *The Economy of Cities*.
73.

and skills; creativity in itself is not enough. Second, most creative initiatives by humans fail or end up being vacuous, either as a result of being badly conceived in the first place or because they do not achieve a sufficient level of social diffusion or acceptance. Furthermore, and as Schumpeter emphasizes, one might think oneself very creative, but his or her competitors could be even more creative. One must identify and measure just how successful creative proposals are. And it is not enough to merely say that this person is creative, even though he or she might fail. Innovation processes are much more complex and difficult than their simple inventive or creative phases. Moreover, as we have seen, many innovations stem from imitation and the transferring of an innovation from one field or area to another – without any creativity whatsoever and simply by means of knowledge transfer. Third, innovations are destructive as well as creative, as Schumpeter knew very well. In the case of innovation studies, any creativity index should be accompanied by a destructiveness index. Florida does not do the latter. His creativity indexes are not very reliable, for all the popular success they may have had.

Innovation in the Public Sector

By means of an introduction to this section, I will return once more to the *Oslo Manual*, which states that innovation is also important for the public sector. Therefore the authors of the manual, which constitutes a canonical document for the dominant paradigm in innovation according to which it is companies that innovate, acknowledged in 2005 that innovation could stem from the public sector and not just from markets. Since the publication of this latest edition of the manual, the OECD and other agencies and institutions involved in studying innovation have begun to think more about this new form of innovation, which had until that time been little studied. In early 2010, the OECD defined a new strategy for innovation studies and policies, recognizing that there are many forms of innovation besides that based on R&D.[74] In a subsequent

74. In the very foreword to this document, Angel Gurría, the OECD secretary-general, alludes to a more expanded range of innovation studies, beyond R&D alone: "The OECD Innovation Strategy also delivers a new set of indicators, clearly showing that it goes well beyond R&D. This work

report, *Measuring Innovation* (also published in 2010), the OECD defines its priority actions, which include a plan to measure different forms of innovation. Action 3 refers to the public sector. Here, the OECD seeks to "recognize the role of innovation in the public sector and promote its measurement."[75] Later in the same publication, it is emphasized that measuring this form of innovation is not easy, and "because the concept of "public sector" encompasses very different organisational units (e.g., the public administration, the health sector, the education sector), it may be necessary to develop new concepts, such as innovation in education, and different metrics to encompass the public welfare aspects of innovation."[76]

Thus, the OECD itself altered its conceptual framework after 2009,[77] extending it from the private sector to now include the public sector as well. In taking this step, it points out that new instruments of analysis and measurement must be created. This is a particularly important kind of hidden innovation because up to this point only innovation processes in companies had been analyzed, and now they were to be studied in different public administrations. Setting aside, in large measure, formerly held neo-Schumpeterian ideas, the OECD finally acknowledges that it is not just companies that innovate but that there are other political and social agents (governments, public administrations, civil society, and the like) that also generate innovations, even though they are not designed to increase competitiveness in markets but instead to promote other social values: "the public sector has multiple objectives, including innovation for social goals."[78] One might conclude, then, that there are forms of innovation based on generating social and not just economic value. Some innovations produced in the public sector are of this kind. While the public sector does take into account the economic dimensions of innovation processes, it also pays attention to other values and objectives. This reworked

---

reflects the plurality of innovation actors and processes and outlines a measurement agenda for furthering progress in this area." See OECD, *The OECD Innovation Strategy*, 4.

75. OECD, *Measuring Innovation*, 14.

76. Ibid.

77. See OECD, *New Nature of Innovation*, 2009.

78. OECD, *Measuring Innovation*, 14.

and extended conceptual framework implies a certain degree of acceptance for social innovation and, in any event, considerably extends the range of innovation studies and policies in bringing several forms of hidden innovation to the table. This important conceptual transformation on the part of the OECD had an immediate impact on the methodology used in questionnaires and surveys, as well as in the emergence of other kinds of innovation studies, as we will see below.

Anthony Arundel and Hugo Hollanders, two of the people involved in designing the European Innovation Scoreboard (EIS), underscore the importance of innovation studies in the public sector. Focusing on Europe, they discuss those that have appeared to date:[79] the pioneering 2001 work of Sandford Borins on public sector innovation in the United States and several Commonwealth countries;[80] the 2011 Nesta pilot survey of innovation in the British health care and local government sectors;[81] the 2010–11 MEPIN project to develop a survey questionnaire for innovation in the public sector in Nordic countries;[82] and lastly, the European Commission's 2011 Innobarometer (IB) survey on the public sector across Europe.

Comparing the latter three studies, Arundel and Hollanders make the following observations:

> In comparison with the private sector, there has been little rigorous empirical research on how innovation occurs in the public sector, with most research based on case studies or very small-scale surveys. There are exceptions, particularly for health care services and education and for the adoption of specific technologies, but until recently there have been no large-scale surveys of how local, regional and national public sector agencies that are responsible for a range of government services innovate.
>
> Up until the early 2000s, the largest survey of public sector innovation was probably the study by Borins (2001), who surveyed 300 public agencies in the United States and several Commonwealth countries. Since then, three separate surveys of public sector innovation were implemented in 2010. The smallest is

79. Arundel and Hollanders, *A Taxonomy of Innovation*, 4.
80. Borins, "Encouraging Innovation in the Public Sector."
81. Hughes, Moore, and Kataria, *Innovation in Public Sector Organisations*.
82. See Bloch, *Measuring Innovation in the Nordic Countries*.

> the NESTA pilot survey of health organisations and local government organisations in the UK, which obtained 175 responses (Hughes et al., 2011). The MEPIN project developed a survey questionnaire for innovation in the public sector that adapted the third revision of the Oslo Manual (2005) guidelines for how to measure innovation in the private sector. Pilot versions of the MEPIN questionnaire were implemented between May and October of 2010 in Denmark, Finland, Iceland, Norway and Sweden, although the type of public agency that received the questionnaire varied in each country. Four countries included hospitals, and two also included secondary schools. Responses were obtained from 2,012 public sector organisations.
>
> The third and largest study is the European Commission's Innobarometer (IB) Survey, which is the data source for this paper. The survey was limited to sectors that are dominated by the public sector (with the exclusion of health and educational services) and obtained 3,699 responses from public organisations. All 27 European Union countries plus Norway and Switzerland were included in the survey. The IB questionnaire was developed by MERIT in collaboration with Gallup (the survey organisation) and the European Commission.[83]

Although different in certain aspects, the three above-mentioned initiatives – that of Nesta, MEPIN, and the IB survey – coincide on one important point: a definition of innovation in the public sector that is very similar, having been based on that of the *Oslo Manual*. For that reason, the three studies expand the OECD's conceptual framework, but do not alter it in any profound way. To date, there has been no change in the dominant paradigm in innovation studies.

The Nesta pilot survey, which began in 2009 and whose results were published in 2011, formed part of a wider plan to create an innovation index for the United Kingdom. After acknowledging the difficulties in measuring public sector innovation, and especially outputs, its authors, Alastair Hughes, Kyla Moore, and Nimesh Kataria, focus on health care and local government. Their definition of innovation coincides to a great extent with that of the *Oslo Manual*: "An innovation is the implementation of a new or significantly improved product (goods or service), or process, a new mar-

83. Arundel and Hollanders, *A Taxonomy of Innovation*, 4.

keting method, or a new organisational method in business practices, workplace organisation or external relations."[84] They then go to recognize significant differences between innovation in the private and public sectors:

> a. Differences in how value is defined. Innovation in the public sector is assessed through impact on a range of social value as well as economic value indicators.
> b. Differences in the systems in which organisations operate. The framework reflects that public sector organisations operate in a range of different systems and assesses the impact of the system conditions on innovation in organisations.[85]

For my part, I would underscore the first difference, given that it overlaps significantly with the wider axiological perspective I propose.

As regards the MEPIN project, it offers a new element that is methodologically very important: the comparability of data among different countries, something that Nesta did not attempt in its survey. "The objective of the Nordic research project *Measuring innovation in the public sector in the Nordic countries (MEPIN)*," states Carter Bloch, "is to develop a measurement framework for collecting internationally comparable data on innovation in the public sector, which both will contribute to our understanding of what public sector innovation is and how public sector organisations innovate and will develop metrics for use in promoting public sector innovation."[86] Yet the pilot questionnaire, with the subheading *The Copenhagen Manual*, used by these Nordic countries – Denmark, Finland, Sweden, Norway, and Iceland – does not especially expand our understanding of the outputs generated by innovation processes in the public sector.[87]

---

84. Hughes, Moore, and Kataria, *Innovation in Public Sector Organisations*, 5.
85. Ibid.
86. Bloch, *Measuring Innovation in the Nordic Countries*, 4.
87. The subjects covered by the MEPIN questionnaire are the following: "Innovations; Innovation activities and expenditures; The objectives of innovations; Information channels for innovation activities; Innovation cooperation; External funding for innovation; Innovative procurement practices; Driving forces of innovation; Innovation strategy, management and competences; Barriers to innovation." Ibid., iv.

This aspect was addressed by the EU IB survey in 2011, which includes five questions about the positive benefits of innovation in the service sector, four on organizational and process innovation, and another four on innovation in social communication. It also contains seven questions aimed at inquiring into obstacles to innovation in the public sector, which is an interesting contribution that was also part of the MEPIN pilot questionnaire (although with less emphasis).[88] It is worth pointing out that the fourth type of innovation distinguished by the *Oslo Manual* in the private and business sector – marketing innovation – undergoes a name change when it comes to the public sector and becomes "communication innovation." Indeed, the MEPIN project uses the same terminology in its project. Therefore, when analyzing the public sector one comes across a new variant of innovation – communication innovation – that is broader in scope than marketing innovation and much more interesting from the point of view of social innovation.[89]

These three European studies have, then, opened up a new field for innovation studies in general. One hopes that similar studies will be carried out in the future in other countries across the globe, either through the OECD, the United Nations, or any other international organization. The 2011 IB survey, drawn up by the Maastricht Economic and Social Research Institute on Innovation and Technology (UNU-MERIT), is at present the most extensive study of its kind at an international level on innovation

88. Arundel and Hollanders, both members of the Maastricht Economic and Social Research Institute on Innovation and Technology (UNU-MERIT), highlight the importance of studying barriers to innovation in the public sector and not just among innovative agents: "Obstacles or barriers to innovation in the public sector have attracted a lot of attention, possibly because of a belief that the obstacles faced by public agencies differ from those in the private sector. Barriers that are possibly more important in the public sector include risk aversion, resistance to change on the part of users, regulatory requirements that limit change, a lack of resources, and professional and management resistance (possibly part of a risk averse culture)." Arundel and Hollanders, *A Taxonomy of Innovation*, 6.

89. In *Measuring Public Innovation in Nordic Countries*, Bugge, Mortensen, and Bloch also advance the idea of using the concept of communication innovation instead of marketing innovation.

in the public sector, in that it covers twenty-nine countries. We will have to see if other OECD countries as well as non-OECD countries undertake similar surveys in the coming years, which is foreseeable.

From my own conceptual perspective, the three aforementioned studies suffer from the drawback of remaining too faithful to the dominant paradigm in innovation, the *Oslo Manual*. They do not encourage a paradigm shift but rather broadening the field in which the dominant paradigm is applied. And this leads to conceptual and methodological problems, of which I will highlight two that seem to me especially significant.

First, the people charged with carrying out the IB survey interviewed representatives of 4,063 organizations in European public administrations, but when it came to selecting people to interview, they always chose the managing directors or strategic heads of these institutions. As a result, one of the key dogmas of the dominant paradigm and heir to the RDI linear model was retained: namely that innovation flows from top to bottom, and that is why company or institutional directors must be surveyed. As we will see below, challenging this dogma is vital when it comes to promoting an alternative paradigm in innovation studies. Innovation processes and their results must be analyzed from various points in the value network, not just from a supposedly higher point (management).

Second, he fields chosen to study the public sector should be expanded. The public sector is always conceived in terms of education, health care, and certain dimensions of public administration, but nothing else. There are important subsectors in the public sector that have not been examined, and that are not even taken into account by canonical studies of innovation. This is the case of the judiciary, police, and armed forces, to mention just three significant subsectors for the res publica. And there are others: for example, public transport, sport (insofar as it is promoted by public administrations), and the cultural sector, which I discussed above. Furthermore, there are several sectors in which the public and private interact closely, for example in the emergency services. However, only private companies tend to be studied in this sector.

Examining these hybrid sectors would be one way of continuing to expand innovation studies, revealing new forms of hidden innovation. Indeed, as the MEPIN project states:

> The public sector has traditionally been viewed as being radically different than the private sector in terms of innovation, with the public sector often seen as a regulatory framework for innovation in the private sector, and as a passive recipient of innovations from the private sector. However, in recent years, public sector innovation has been increasingly regarded as a central factor to sustain a high level of public services for citizens and businesses, as well as addressing social challenges and improving welfare. Public sector innovation may have considerable effect not only on the quality and efficiency of public services itself, but also may influence the private sector's ability to innovate.[90]

These considerations deserve to be studied in detail in the diverse public subsectors rather than here at this time. In order to conclude this section I will offer some brief thoughts on innovation processes in military environments, given the historical importance of this sector for the emergence of technoscience. This form of innovation has never been analyzed by the *Oslo Manual*, nor is it typically discussed by academic studies on innovation, despite the fact that military innovation is one of the best examples of technological innovation. The atomic bombs in Hiroshima and Nagasaki and, more generally, the Manhattan Project constitute a canonical example of military innovation generated by scientific research and technological and industrial developments (R&D). The same could be said for many other technoscientific advances in the twentieth century, many of which originated in US military R&D. It is surprising, then, that academic studies of innovation hardly ever address military innovation processes, nor are they ever analyzed empirically. To my knowledge, the questionnaires of the *Oslo Manual*, Nesta, the MEPIN project, and the EU IB have never been applied to a military institution, and possibly nor to security agencies or organizations. It is a significant anomaly that makes up a new kind of hidden innovation. Despite the fact that, in principle, military innovation fits both the linear RDI model and the technological, manufacturing, and product innovation mod-

90. Bloch, *Measuring Innovation in the Nordic Countries*, 3.

el quite well, it has not been studied with the empirical analysis methodologies I am talking about here. In my view, this is not actually hidden innovation, given that the existing indicator systems allow one to analyze it without any major problems. It is, rather, intentionally "covered-up innovation" that appears to be outside the reach of academic studies on innovation for reasons one would have to look into.

In order to understand this anomaly, at least in part, I will dedicate a few brief paragraphs to discussing why military innovation processes – which do exist and which are particularly important in the United States as well as other countries – may have been excluded from broader studies of the phenomenon. Military innovation originates, at least in principle, in the public sector, which finances the R&D that generates this innovation. It would appear that studies of innovation by the OECD, Eurostat, and other organizations use extreme caution when it comes to examining this form of innovation. Here I will briefly touch on military innovation, although I also think it important that other forms of covered-up innovation in the public sector are also examined.

As regards military innovation, one might think that the important thing is to generate new weapons, whether for attack or defense. If this were the case, it would be product innovation, much of which stems from R&D as was the case in the examples of chemical weapons (World War I), nuclear weapons (World War II), and more recently, biological weapons. In order to think up new weapons, there must be new scientific knowledge and new technological developments, besides industrial manufacturing that makes these weapons operative for the military in such a way that it possesses better weapons than the enemy. Technological reductionism tends to believe that military victory is the direct result of such technological innovation, at least in the case of twentieth-century wars, which were carried out with industrial and information weapons based on scientific and technological knowledge.

However, the theories available on military innovation are more complex and polished than the linear model version outlined in the previous paragraph. In a work published by the Rand Corporation – one of the organizations that has done most to think about the nature of contemporary warfare and armed forces – and authored by Jeffrey A. Isaacson, Christopher Lane, and John Arquilla, it is argued from the outset that military innovation processes do not just refer to new products but also to new processes, to new forms of internal organization, and (in the terminology of the MEPIN project and the IB) to new communication forms – both internal (the telephone, walkie-talkie, Internet, mobile telephone, and so on) and external (the "red telephone" or Moscow–Washington hotline) to the military. In the words of the authors, "Technology alone does not determine military effectiveness. Specialized doctrine, tactics, training and support are generally required to integrate, or absorb, technology into a military organization."[91]

Supplying a military organization with weapons is of little use if the soldiers do not then know how to use these weapons and have not been trained to do so. A new weapon is not just a question of hardware, but also and above all of the ability and skill needed to use it. Therefore, conceiving of military innovation as a technological innovation process alone, which takes place in scientific and engineering laboratories in which new weapons are generated and made, is a grave error. The appropriation of new technologies by soldiers and their superiors is just as important as innovation, strictly speaking, in such a way that this is also a case of (at least) technosocial innovation. In fact, military organizations spend a lot of time on military training, which is done with weapons but whose objective is to generate new abilities in the people being trained, including the capacity for organization and communication as well as new protocols for action in specific circumstances (process innovation). A general problem, then, emerges, to which I will return later: *the appropriation of new technologies on the part of their users* (in this case, soldiers). If this step is unsuccessful, a military innovation process fails and the new weapons are useless because of a lack of users with the corresponding know-how.

Consequently, there are various forms of innovation in military organizations besides product innovation: "Improving effectiveness in these militaries cannot be achieved by simply buying hardware: in general, some level of innovation is required – e.g.,

91. Isaacson, Lane, and Arquilla, *Predicting Military Innovation*, 1.

reorganizing command structures, introducing new doctrine and tactics, modifying training techniques, and improving support."[92] One could say much the same for another important field in innovation studies, that of sports, in which training, tactics, and the development of abilities usually contribute forms of innovation that are decisive when it comes to success or failure in sporting competition.[93] In the case of military conflicts, new weapons usually play an important role but are not always decisive, nor do they necessarily predetermine the outcome of the combat. In short, in the military sector, too, there are various kinds of innovation and not just that based on R&D. This is the main consequence that must be drawn from this brief commentary on military innovation. In order to end this aside, I will cite the definition of innovation used by Isaacson, Lane, and Arquilla to describe this public subsector: "We define military innovation as follows: for a specific military, innovation is manifested by the development of new warfighting concepts and/or new means of integrating technology; new means of integrating technology might include revised doctrine, tactics, training or support."[94]

This definition has little to do with that of the *Oslo Manual* or others that derive from it used to study innovation in the public sector. Military innovation studies have followed a different trajectory to that of economic innovation studies, despite the fact that the TPP model would appear to be very pertinent to the military sector. Theoreticians of military innovation have developed their own conceptual frameworks. In fact, the previously mentioned study for the Rand Corporation by Isaacson, Lane, and Arquilla aims to provide a set of indicators in order to detect military innovation – indicators that have little in common with the innovation indicators of the OECD and Eurostat.

I would conclude, therefore, that the indicators and measurement methods associated with the *Oslo Manual* are only valid for very specific forms of innovation. Many other types of innovation,

92. Ibid., 2.

93. In military practice, as in police and sports practice, it is often a case of generating automated or tacit knowledge. This dimension deserves to be studied in more detail.

94. Isaacson, Lane, and Arquilla, *Predicting Military Innovation*, 8.

however, are outside the field of analysis of the *Oslo Manual* indicators and should therefore be considered as hidden innovation in the technical sense of the term: namely, those innovation processes that are not analyzed according to the indicator systems of the OECD and Eurostat.

User Innovation

In 1988, Eric von Hippel published a work in which, on the basis of several empirical studies, he states that there are various sources of innovation: "In some fields, innovation users develop most innovations. In others, suppliers of innovation-related components and materials are the typical sources of innovation. In still other fields, conventional wisdom holds and product manufacturers are indeed the typical innovators."[95] In order to reach these conclusions, which go completely against the linear model of innovation, von Hippel and his colleagues carried out studies of functional sources of innovation in different manufacturing sectors: scientific instruments, semiconductor and printed circuit boards, pultrusion, tractor shovels, engineering plastics, plastics additives, industrial gasses, thermoplastics, and wire termination equipment.[96] They did not start from an a priori definition of innovation: instead, they asked different business people and professionals what important innovations had been produced in their sector in recent years, selected a list of case studies, and from that list inquired company by company about the origin of each of these innovations. This empirical study led von Hippel to conclude that they would have to replace "the manufacturer-as-innovator assumption with a view of the innovation process as predictably distributed across users, manufacturers, suppliers, and others."[97]

Later the functional chain concept is introduced, with von Hippel differentiating at least four different links in a process of production and consumption: manufacturers, suppliers, distributors, and consumers (or users). Von Hippel openly admits that there can be more links in this chain such as insurers, for example,

95. Von Hippel, *The Sources of Innovation*, 3. Von Hippel had already published several articles alluding to this thesis through the 1980s.
96. Ibid., 4.
97. Ibid., 3.

which in some sectors have also been sources of innovation. And his general idea is that in every functional chain there are various potential sources of innovation, and that this issue can only be studied empirically, according to the respective context and sector. He makes no attempt to define innovation at any point and is very clear that the meaning of this word varies according to different sectors and the case studies analyzed. This empiricist conception of innovation has its roots in the work of Friedrich Hayek, for whom only market processes could reveal which innovations were better than others.[98]

Still following an empiricist methodology, von Hippel proposes a conceptual framework in order to analyze innovation processes, and in particular to locate different sources of innovation. Manufacturers (producers), providers (suppliers, distributors), and users (consumers, customers) are potential sources of innovation in different economic sectors in that they are the functional classes in a chain of production and consumption. This notion implied an alternative to the linear model in the late 1980s, but only contributed an academic debate and did not manage to inspire a change in innovation policies.[99] Yet twenty years later these same ideas gained increasingly widespread acceptance, especially in Europe, where user innovation has become a key issue. For my part, I will take on board von Hippel's proposal and attempt to formalize and develop it, with the aim of having an alternative conceptual framework at my disposal in order to examine innovation processes. If only for its heuristic function, distinguishing different functional classes in a chain of production and consumption is of great interest. Nevertheless, the notion of a functional chain was ultimately superseded by that of a value chain, which as noted was introduced by Porter in the 1980s.[100] This latter idea allows one to analyze a

98. F.A. Hayek, an economist of the Austrian school, conceived of market competition as a means of discovery and innovation.

99. Prior to von Hippel, in the mid-1970s Christopher Freeman had already pointed out that a lot of innovation originated among users, but his ideas had little impact at that time. See Freeman, *The Economics of Industrial Innovation*. In contrast, von Hippel, has made a significant contribution to innovation policies in the United Kingdom and Europe, especially those implemented after 2005.

100. Porter, *Competitive Advantage*.

company by differentiating its constituent parts with the aim of calculating the value that each of these components generates.

Later, Porter himself broadened this idea tone of value systems, in which the value chains of a company are linked to the value chains of other companies. For me, there are various types of value, and especially social values. Therefore one should not just talk in terms of economic value chains, but also social value chains, for example in the public sector or in the third sector. The notion that seems most appropriate to me when it comes to innovation studies is that of a value system, in which economic values can be combined with social values and values of other kinds.

As a first step in that direction, I will reinterpret von Hippel by using the notion of *innovation systems* instead of functional chains. This eliminates the linear notion associated with the term "chain" and allows one to identify different points and subsystems of innovation. I will, then, revisit the notion of an innovation system (along the lines of Lundvall and Nelson) that operates at different levels (local, regional, national, supranational, global, and so forth). Yet when it comes to analyzing an innovation system I will be more interested in the functions and relations that exist between its points than in the points themselves. This implies using a pluralist and relational model of innovation in which there are several sources of innovation, with the emergence of innovation processes depending on the relations among these elements (or points) in the system. Connections and relations among different elements and parts of the system are, then, decisive when it comes to facilitating the emergence of innovation. Such innovation does not stem from one point alone but rather from the interrelations among two or three points: this is what gives the model a relational and not just pluralist dimension.

Initially, following von Hippel, I will distinguish four points: manufacturers (producers), suppliers, distributors, and users (consumers). The relations among these four points (M, S, D, and U in the figure below) are multiple and must be studied empirically. Moreover, these relations are not unidirectional, as in the linear model, but bidirectional. I thus propose this tetragonal model:

The distinction between suppliers and distributors is not al-

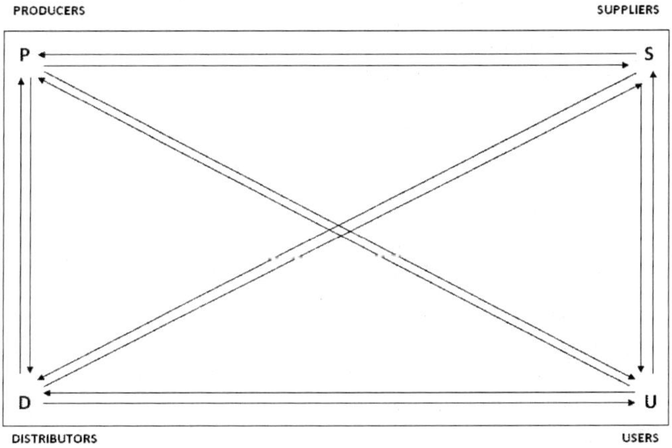

Figure 2: Tetragonal Model of Innovation Systems .

ways clear, although in some cases it is. In general, there are always intermediary points between manufacturers or producers and users or consumers. Obviously, this quadrangular model can easily be extended to form a polygonal model, with *n* vertices. The important thing is to analyze all existing relations among the four points (or the *n* points), explaining whether they are bidirectional. Ternary or *n*-ary relations, in which various points in the innovation system intervene, are of particular interest. What is important is that von Hippel's idea can be broadened, giving rise to a formal analytical instrument that might be of use to different economic sectors and, more generally, for different kinds of value that may be of interest in innovation studies.

I will also make a dissection between proposed innovations and effective innovations. Formally speaking, proposed innovations or p_i can be characterized as functions that go from one point to another (or to others) and that are received by those other points as innovations, to be subsequently evaluated positively or negatively. In the most typical cases, relations among the points in the system are commercial in nature; in other words, one point acquires or buys a proposed innovation (p_i) from another. But there are many different kinds of relations among the points in an innovation system such as collaborative relations or knowledge transfer relations. It is also typical for one point to offer or publicize an innovation (p_i) to another and for the latter to receive it in one way or another (with mistrust, skepticism, enthusiasm, interest, and so on), and all this before it arrives on the market. A proposed innovation may originate in specific relational environments and then be diffused throughout the whole system. Commercial relationships are not the only forms that can be used to analyze innovation processes. Many innovations improve supply, distribution, and/or use, and not just the production cost. The formal and informal transfer of knowledge also leads to proposed innovations (p_i). In short, innovation processes must be analyzed by not only their final outcome in the corresponding market but throughout their processes as well: if not, the effective innovation source is not detected. The emergence of an innovation is a complex interaction among different points in the system whose main result consists of acceptance of the proposed innovation and its subsequent adoption on the part of at least one of the other points. The most significant innovations affect all the points in the system, as we will see later.

This formal model can be implemented in different ways: for example, by weighing up the aforementioned level of acceptance according to the price paid for p_i, in line with the time used up by the innovation, or in relation to positive or negative attitudes that may be detected toward p_i. In proceeding this way, it is possible to examine the culture of innovation from within its own dynamics and not just via the ultimate results obtained. The important thing is that the pluralist systemic model can represent different types of relations among the points and not just commercial relations. In the case of social innovation, one would be analyzing social relations. And obviously, not all those relations would be measurable on the metric scale.

Von Hippel makes other important points: First, *the same agent can adopt several roles*. An automobile firm, for example, is a manufacturer in its factories but also a supplier of the cars it makes, either by sea, train, or trailer. It may even choose, as is typical, to have its own distribution network, normally through licensed car dealerships or franchises. However, it is also a user of multiple components that other firms manufacture for it (engines, tires, batteries, seats, and so on). It can even increase its presence in the system of production and consumption by creating a network of automotive repair shops, offering financing and insurance deals, proposing leasing formulas, and so forth. The presence of the same firm at multiple points in the system like this is quite typical in the industrial sector, but also occurs in the service sector. One of the hypotheses I defend in this regard is that an agent that embraces many roles is more likely to be an innovator than another that always remains in the same role; this is due to the fact that the former has more knowledge of the innovation system as a whole.

Second, von Hippel also contends that communication among the different functional points favors innovation because it makes knowledge transfer possible among these different points in the system. Whether it is a question of the advantages or disadvantages of a product or service, (formal or informal) communication channels among the points in the system favor innovation. As examples, von Hippel mentions scientific publications on the benefits and defects of different instruments of measurement, observation, and experimentation, patents, visits to factories, and informal channels of communication among forms in the same sector. I will examine the first example briefly. A scientist produces knowledge, for example in the laboratory, and also distributes this knowledge by means of journal articles and presentations at conferences. However, the same scientist is also a user of apparatuses and instruments whose benefits and defects he or she may evaluate on the basis of being an expert in their use. Routine use of an artifact generates knowledge, and the issue consists of transferring this knowledge to the rest of the system. The custom of scientific communities is not just to publish scientific articles that offer data but also others that warn of defects or errors in measurement instruments; this constitutes one of the main sources of innovation in the field of scientific instruments, which the empirical studies of von Hippel's research team confirm. The existence of channels of communication among manufacturers, suppliers, distributers, and users of scientific instruments, guaranteed by protocols, reports, and scientific publications, strengthens innovations in this sector.

Third, problems and conflicts among functional points can also be sources of innovation, for example as a result of complaints about defects in the functioning of another point in the system or a particular artifact or service. Claims and complaints by suppliers, distributors, and users are a source of knowledge and, hence, potential innovations. The plurality of innovation sources includes negative as well as positive signal relations among the points. Indeed, this is one of the most typical mechanisms for innovating: detecting the functional defects of something. Paraphrasing Karl Popper, I will advocate a kind of *methodological falsification* in order to analyze innovation processes, and for that reason I am interested in barriers or obstacles to innovation, failure in innovation, and the difficulties involved in innovating, and not just in innovating agents, success in innovation, and the fostering of innovation.

Fourth, von Hippel advances an idea that he has returned to on several occasions throughout his career: lead users. In 1986 his definition was the following: "Lead users are users whose present strong needs will become general in a marketplace months or years in the future."[101] Therefore, these are users who think about what they are using and who are capable of imagining improvements and implementations, and even carrying these out by themselves. A typical example of this, which von Hippel has cited on more than one occasion, is that of free software developers and, more generally, hackers. In his 2005 work *Democratizing Innovation*, lead users become one of the main ideas of his overall theory:

> When I say that innovation is being democratized, I mean that users of products and services – both firms and individual consumers – are increasingly able to innovate for themselves. User-centered innovation processes offer great advantages over the manufacturer-centric innovation development systems that have been the mainstay of commerce for hundreds of years. Users that innovate can develop exactly what they want, rather than relying on manufacturers to act as their (often very imperfect) agents.[102]

101. Von Hippel, "Lead Users," 791.
102. Von Hippel, *Democratizing Innovation*, 1.

In general, these lead users are specific people, but they can also form groups if they bring themselves to sharing the knowledge they have with other users, in which case innovation communities emerge. As von Hippel goes on to state, "Moreover, individual users do not have to develop everything they need on their own: they can benefit from innovations developed and freely shared by others."[103] And, "Innovation communities can increase the speed and effectiveness with which users and also manufacturers can develop and test and diffuse their innovations. They also can greatly increase the ease with which innovators can build larger systems from interlinkable modules created by community participants."[104]

The knowledge that users acquire increases with experience on the basis of the repeated use of specific products or services, and thinking about them in accordance with how much they use them. In this way, problems, deficiencies, and malfunctions are detected, and occasionally, the means to correct such problems are voluntarily offered by the users, whether in the case of products or services. Insofar as these ideas are mutually communicated, tacit user knowledge is codified and, in some cases, generates new ideas that in turn give rise to innovations. This process is much more typical than it may appear to defenders of the linear model: following numerous empirical studies, von Hippel argues that at least 25 percent of all innovations come from users, and this based on different economic sectors. There are certain sectors, though, in which user innovation is more typical. And this leads von Hippel to propose a new innovation studies model, namely the user-centered model:

> The user-centered innovation process just illustrated is in sharp contrast to the traditional model, in which products and services are developed by manufacturers in a closed way, the manufacturers using patents, copyrights, and other protections to prevent imitators from free riding on their innovation investments. In this traditional model, a user's only role is to have needs, which manufacturers then identify and fill by designing and producing new products. The manufacturer-centric model does fit some fields and conditions. However, a growing body of empirical work shows that users are the first to develop many

103. Ibid.
104. Ibid., 11.

and perhaps most new industrial and consumer products.[105]

In my opinion, these arguments are very important for innovation studies because they contribute a new paradigm, and not simply an extension of the dominant paradigm: First, because they break the vertical model according to which innovation stems from R&D and is carried out by entrepreneurs. Second, because they call into question the idea that innovation only emerges to meet needs, which in the case of social innovation means social needs. This is not the case, as von Hippel maintains, because the solving of problems and even satisfying of desires are also driving forces of innovation, in particular social innovation. Third, because von Hippel contributes a communal dimension to innovation studies and does so without renouncing the idea that it is individuals who innovate. The linear model and even many versions of the systemic model of innovation underscore the importance of knowledge transfer, for example from universities to companies. Innovation communities have another kind of structure because they share knowledge, which is very different than transferring it. Even the notion of "transfer" has a connotation of economism that the word "sharing" does not because the latter is much more general in nature and can be applied to anyone. Needs and knowledge can be shared, but so too can projects, activities, dreams, feelings, amusements, and, obviously, money.

For these three reasons, to which I will add others in the following paragraphs and the next chapter, von Hippel's proposals call into question various fundamental principles of the paradigm associated with the *Oslo Manual.* For that reason, they make up an alternative theoretical framework for innovation studies and, where appropriate, for innovation policies as well. In recent years, other authors have come to notice this as well and have embraced some of the ideas associated with von Hippel. He typically speaks about distributed innovation because it is not just users that innovate; innovation also stems from other points in the value system. Among the many examples that could be offered on the recent influence of von Hippel's ideas, I will mention two: one academic, the other institutional.

105. Ibid., 2.

The first refers to a researcher at the MERIT institute who has carried out studies on the European Union of the kind we have already come across in the case of Arundel and Hollanders. His name is Fred Gault, and in March 2011 he published a working paper in which he demonstrates that, despite its successive conceptual additions, user innovation is still excluded from studies of innovation based on the *Oslo Manual*.[106] In the same document, Gault proposes modifying the definitions used by the manual in such a way that its conceptual framework is broadened to incorporate user innovation. It is an interesting idea, and he introduces a new way of studying hidden innovation: namely, by modifying the conceptual and methodological tools that are used in innovation studies, whether they are definitions, proposed applications, measurement methodologies, indicator systems, or other tools. Instead of proposing an alternative paradigm, then, Gault (and others) continue with the strategy of "patching up" the *Oslo Manual*. In order to do this, they partly alter the conceptual framework of the document and introduce new analytical instruments with the goal of examining other forms of innovation, besides those for which the manual was designed. This is the same strategy adopted by the OECD and Eurostat through three editions of the manual, and it has been applied to studies of innovation in the public sector, as we saw previously. As more and more examples of hidden innovation emerge, so the OECD extends its conceptual framework and introduces complementary measurement indicators and methods.

For my part, I think a strategy like this will be ultimately unsuccessful in the case of user innovation. In order to argue my case, I will discuss in some detail Gault's proposals, in part also because they have been published precisely to stimulate discussion on this issue. Nevertheless, I will also bear in mind von Hippel's previous theses as well as other recent European documents on user innovation.

As regards Gault's working paper, at the outset it highlights a significant deficiency in the *Oslo Manual* that I already referred to in the previous chapter:

> According to the Oslo Manual innovation is a market phenom-

106. Gault, *User Innovation and the Market*.

> enon. A new good or service, or process, is an innovation only if it is connected to the market. This connection works for firm based innovation as the firm exists to sell products (goods or services) to the market. As part of its business strategy, a firm will launch a new product on to the market or use a new process to move products to market. Consumers are another matter. They do not exist to sell products to the market.[107]

There are innovations aimed at markets, but the *Oslo Manual* only discusses these. Therefore other forms of innovation remain outside its gaze precisely because it is restricted to the notion of companies as the only innovative agents and markets as the sole social spaces in which an innovative proposal can be successful. The issue here, then, is hidden innovation, much of which is very significant. I am not just referring to innovation that originates in the public sector or civil society, but also to the many innovation processes that take place at the "margins" of the private sector, to put it one way. A typical case of this would be those users that, on their own initiative, modify some product they have acquired that, for whatever reason, does not adequately satisfy their needs or demands. In doing so, they generate new products that are of use to them. Occasionally, those products are given freely to other users, and as a result new products emerge that function outside the limits of the market, in exchange-based circuits among user communities. An example of this, as noted, is free software, which has been studied thoroughly by both von Hippel and Gault.[108] There are also numerous examples of new products and services created by users, which are improvements on their previous incarnations and end up being adopted or bought by companies. Indeed, according to von Hippel (and a team of other authors), "Many successful products in the market were initially developed by users, and eventually adopted and commercialised by producers."[109]

107. Ibid., 1.

108. Gault, and von Hippel, *The Prevalence of User Innovation and Free Innovation Transfers*. This working paper clearly states the need to create specific indicators for user innovation, arguing, "Statistical indicators used in official surveys of innovation activities have not addressed this new understanding of the central role of users in the innovation process. New indicators must be created to provide a clearer picture" (3).

109. Flowers et al., *Measuring User Innovation in the UK*, 4.

Innovation comes from users, although it can later be developed by commercial profit-making firms and not just user communities. There is no antimony between producers and users here; rather, they are two points in a value system in which innovation might originate at any point. It is important to stress that both Gault and von Hippel (and their colleagues) do not just refer to end users or consumers but also to firms that use products or services created by other firms. In *Democratizing Innovation*, von Hippel defines these two forms of innovation in the following way: "Users, as the term will be used in this book, are firms or individual consumers that expect to benefit from using a product or a service. In contrast, manufacturers expect to benefit from selling a product or a service."[110] Therefore a firm can be the user of one product and the producer of another. In fact, this is the most typical situation because all firms have their own suppliers of material or services and are therefore users of these. A firm can be a frequent user of a specific product created by another firm without any hint of industrial espionage, and having tested the product's potential advantages and disadvantages, the firm might choose to improve on it or even make it itself, generating an innovation without infringing on patent or intellectual property laws. This is quite typical in the service sector, especially if they are provided by public institutions. The frequent use of a specific service generates knowledge about this service and, occasionally, ideas about improving it.

User innovation is also related to Chesbrough's notion of open innovation and can be produced in any economic sector, including certain innovations that stem from R&D. This is not something anecdotal or complementary to the TPP model of the *Oslo Manual*. Many user innovations occur in important sectors of business activity. It is therefore preferable to speak in terms of product, process, organization, or even (a specific) advertising campaign users, rather than clients or consumers. Users can be people but they can also be firms, institutions, or organizations. The key thing is that good communication between users and producers should favor innovation. And this argument is valid not just for users and producers but also for any points in the value system or network. This

110. Von Hippel, *Democratizing Innovation*, 3.

will, in fact, be the key aspect in my own argument: the pluralist and relational model of innovation.

Let us recall von Hippel's definition of users as "firms or individual consumers that expect to benefit from using a product or a service."[111] This is a good starting point, but we should extend it in two directions. First, public institutions and administrations are also users of goods and services, as are organizations in the third sector. Therefore many user innovations can be considered as social innovations in the event that they generate social value. Second, a worker may be considered a user in a particular organization, institution, or company, even if it is just because he or she uses its resources, capital goods, or services. So long as a worker asserts his or her tacit knowledge and transfers this to the company management, he or she becomes a source of innovation that can also be catalogued within the user form of innovation. This form of innovation was originally theorized, as noted, by Nonaka and Takeuchi, but from another perspective;[112] however, it shares a profound similarity with user innovation as understood by von Hippel and Gault. Specifically, both sets of authors are interested in those users that modify products and services through their own initiative, without thinking about selling those innovations later on, which might be the case for any consumer that modifies goods or services, or even develops new kinds of goods or services. In any event, this is hidden innovation because "in official statistics, guided by the Oslo Manual, this consumer activity would not be classified as user innovation unless connected to the market."[113]

With the aim of correcting this deficiency in the *Oslo Manual*, Gault proposes a conceptual change that takes shape in new definitions.[114] After providing ample empirical evidence on the frequency of user innovation in Canada and the United Kingdom, Gault evokes the definition of innovation used in the *Oslo Manual*.[115] He then highlights the fact that the term "implementation"

111. Ibid.

112. Nonaka and Takeuchi, *The Knowledge-Creating Company*.

113. Gault, *User Innovation and the Market*, 1.

114. Ibid., 6.

115. Specifically, the idea that, "An *innovation* is the implementation of a new or significantly improved product (good or service), or process, a new

as used in the manual always implies some kind of link or connection to markets, citing the original text once more: "A common feature of an innovation is that it must have been *implemented*. A new or improved product is implemented when it is introduced on the market."[116] This, actually, is the statement Gault suggests modifying, replacing it with the following: "A common feature of an innovation is that it must have been *implemented*. A new or improved product is implemented when it is *made available to potential users*."[117] Such a change would permit one to consider as an innovation that produced by user communities that share knowledge among themselves and generate new products, but without the aim of selling them on the market. Gault's suggested modification is interesting because it corrects a deficiency of the *Oslo Manual*. Moreover, it can be applied with relative ease to innovation in the public sector, as Gault himself does toward the end of the working paper. Finally, Gault's definition of "implementation" underscores the importance of innovation users in any innovation process, given that the definition of innovation itself now depends on the notion of the user. To my mind, herein lies the most radical shift because the *Oslo Manual* does not attribute any degree of relevance whatsoever to technology users. As I have emphasized on more than one occasion, the *Oslo Manual* model is clearly vertical or "top-down." Including users in the very definition of innovation implies making the concept more horizontal, and much more "bottom-up" than the dominant paradigm is. Consequently, Gault's argument deserves special attention.

Before insisting on this point, I will mention some recent empirical studies that focus on user innovation: a thematic paper by Innometrics for the EU, a Danish study, and a document in which the OECD modifies its strategy regarding innovation studies and appears to accept, albeit timidly, user innovation.

The Innometrics study is based on the Innobarometer (IB) surveys of 2007 and 2009. Authored by Stephen H. Flowers, Tanja

marketing method, or a new organizational method in business practices, workplace organization or external relations," OECD and Eurostat, *Oslo Manual*, §146, 46.

116. Ibid., §150, 47.

117. Gault, *User Innovation and the Market*, 9.

Sinozic, and Parimal Patel, its title, *Prevalence of User Innovation in the EU*, gives a clear indication of its content. It is not misleading, because it offers truly remarkable data on the importance of user innovation in the EU. After making it clear that this is the first study of its kind on the EU as a whole, Flowers, Sinozic, and Patel state that they seek to address two questions: How prevalent is user innovation among a large sample of innovating firms in the EU? And, how are firms engaged in user innovation different from the whole population of innovating firms?[118] In order to answer these questions, they distinguish between three forms of user innovation: user process innovation, user product innovation, and user involvers. User process innovators are "firms that have introduced new or significantly improved *processes* by customizing or modifying those developed by other firms, organizations or individuals." User product innovators are "firms that have introduced new or significantly improved *products* by customizing or modifying products developed by other firms, organizations or individuals." And user involvers are "firms that *involve users* in support of their innovative activities."[119]

The second of these, user product information, is the most studied category, by von Hippel and others. The remaining two categories – user process innovation and user involvers – are new and of great interest to studies of user innovation. Because this study is based exclusively on European firms and the two IB surveys in 2007 and 2009,[120] only cases in which one firm is the user of another are examined. Thus, this study refers to the hard core of business innovation. And it states clearly that this does not just originate in R&D but that it is produced in many other ways, in this case, through interaction among firms, some of which are users of others. This is a new example of hidden innovation that has

118. Flowers, Sinozic, and Patel, *Prevalence of User Innovation in the EU*, 3.
119. Ibid., 10–11.
120. The authors make it clear from the start that the empirical baseline is exclusively made up of firms in that it is based on the 2007 and 2009 IB surveys: "The 2007 survey elicited responses from 5238 firms of which 4400 were self declared innovators. The 2009 survey had 5234 respondents of which 4377 were self declared innovators. The rest of this report focuses on innovating firms only." Ibid., 11.

been detected thanks to a new methodology that differs on key points from that of the *Oslo Manual*. The distinctions and methodologies employed in the study are based on von Hippel's ideas, as Flowers, Sinozic, and Patel openly acknowledge.[121] However, the study itself goes further than the work of von Hippel given that they examine his ideas and apply them more deeply. As they themselves highlight, process innovations have been outside the field of R&D, which has almost always been focused on products. For Flowers, Sinozic, and Patel, user process innovation is clearly one more example of hidden innovation, and they cite the 2007 study carried out by Nesta.[122] They then go on to state that user product innovation is also outside the realm of the linear model. For me, I would say that it is the same case for what they term "user involver" innovation processes, although for different reasons: in order to detect this form of innovation, it is necessary to analyze relations among firms, always remembering that the roles are different when one firm is a user of another.

These conditions are enough to demonstrate the existence of a whole new world to be explored, even within the business sector, beyond the tip of the innovation iceberg that is R&D – even more so if we bear in mind that similar studies could be carried out in the public sector, the CCS, and the third sector, to mention just three areas I discussed previously. Whatever the case, one could conclude that studies of innovation must rework much of the questionnaires they use if they intend to inquire into user innovation. The 2009 Innometrics thematic paper marked a significant conceptual change, especially as it was carried out within the business sector and not in the public sector or in regard to civil society.

While that conceptual and methodological change is important, more important still are the empirical data offered. I will not go into great detail about this and will instead limit myself to referring to the data mentioned in the executive summary of the document, together with some additional information that I find particularly interesting:

> The findings show that a substantial minority of innovative

121. Ibid., 6–7.
122. Ibid., 6. The study referred to is NESTA, *Hidden Innovation*.

> firms in the EU are involved in process and product modification (around 30%), and more than half such firms involve users in support of their innovative activities. User innovation is also more or less evenly spread across industrial sectors and across EU countries.
>
> Large firms are more likely to be involved in all forms of user innovation than small firms. For example 39% of all innovative firms with more than 500 employees are *User Process Innovators*, and in the case of *User Involvers* this rises to 61%.[123]

As a whole, the study demonstrates that over half of all European innovative firms involve their users in the innovation processes they promote: "The results show that of the 4,400 innovative firms in the 2007 IB survey 30.3% are *User Process Innovators*, and 27.7% are *User Product Innovators*. In contrast a much higher proportion of the 4377 innovative firms in the 2009 Survey are *User Involvers*: 53.1%."[124] And another of the study's conclusions is also worth noting: "Firms engaged in user innovation can be classed as 'super-innovators.' Compared to other innovative firms, they are more likely to introduce new products, processes or services. They are also more likely to initiate new organizational methods. Moreover a higher proportion of user innovators carries out both intra and extra mural R&D and applies for patents."[125] It is worth underscoring this latter statement. Those firms that involve users to some degree or another in their innovation processes are much more innovative than those that do not. Therefore user innovation is not a secondary or complementary topic but instead is at the root of the innovative capacity of a company and, one assumes (although this hypothesis would have to be verified empirically), also of administrations in the public sector. In sum, the data presented in the Innometrics study demonstrate that user innovation is a typical practice among innovative European firms, especially in the third form identified by Flowers, Sinozic, and Patel. It is true that this is just an initial study, and it will have to be extended to other sectors and different countries. But in including all the countries of the EU and using such a wide sample of companies, the data seem

123. Ibid., 3.
124. Ibid., 12.
125. Ibid., 3.

to be reliable. From a conceptual point of view, the work opens up new areas of study that will, in my view, imply a radical shift in innovation studies. If such data are obtained from the first time a relational form – that being a firm that uses another firm – has been studied empirically in depth, the outlook for the pluralist and relational model of innovation I advocate is very promising. My model is based on the idea of a plurality of significant points and a multiplicity of relations among these points. I maintain the hypothesis that innovation processes emerge more easily when the different points interrelate and interact with one another through different relational forms. If my conjecture is true, other kinds of relations among firms, and more generally among innovative agents, would have to be studied. And not just from the perspective of knowledge transfer, the canonical point of view in the linear model. The important thing is to share common activities and objectives, regardless of the place that each point occupies in the value chain or network.

Returning to empirical studies, I should mention – although I will not discuss it in detail – a Dutch study published by Jeroen de Jong and Eric von Hippel that is cited by Flowers, Sinozic, and Patel in the following terms:

> While the Statistics Canada surveys focused on all enterprises with more than 10 employees, recent work in the Netherlands (De Jong and von Hippel, 2008) has examined user innovation in SMEs (defined as firms with less than 100 employees). The survey identified firms that either modified, or created from scratch process technologies and, in common with the Statistics Canada survey, collected data on diffusion and expenditures on user innovation.
>
> At a headline level the survey revealed that around 22% of Dutch SMEs were user innovators, (Modifiers: 18%; New Technology Developers: 4%). The Manufacturing sector had the highest proportion of user innovators (Modifiers: 31%; New Technology Developers: 11%), followed by Business Services (Modifiers: 21%; New Technology Developers: 6%), and Farming (Modifiers: 20%; New Technology Developers: 4%). The study also found a strong positive correlation between firm size and the propensity to engage in user innovation (i.e. that larger firms have a higher probability to engage in user innovation

than smaller firms).[126]

As regards the OECD, for the time being it has not published any systematic study on user innovation, but in its latest documents it hints at a certain receptiveness toward this form of innovation, even though it is only considered from a business perspective. The Schumpeterian dogma, according to which companies are the innovative agents par excellence, continues to be at the theoretical heart of the *Oslo Manual* paradigm. Having said that, this model is somewhat more flexible now, because one can read statements such as the following:

> Users, including suppliers and end users, also affect the innovation process by shaping and stimulating market demand for innovation. Studies show that from 10% to 40% of users engage in developing and/or modifying goods and services (von Hippel, 2005). These innovating users are typically "lead users," *i.e.* they are at the leading edge of the market. Innovation by users is often widely distributed rather than concentrated, with the result that innovations are combined and leveraged in so-called innovation communities. In these very direct, informal user-to-user co-operation networks, users help each other to solve problems and innovate.[127]

Innovations are only produced in markets. The OECD continues to affirm this principle, which is constitutive of the dominant paradigm in innovation studies. Although the only remedy its experts have is to cite von Hippel, given his increasing influence in the twenty-first century, they do so with a certain disdain, putting the term lead users in quotation marks and referring to "so-called" innovation communities. Even though, since 2005, the OECD has had to accept several new forms of innovation and has redesigned its conceptual and methodological framework to incorporate these, what it still cannot accept is innovation originating in civil society, for example from an innovation community such as free software communities or many others who base their actions on social values and not on obtaining economic benefit. Economists are strongly influenced by the belief that rationality

126. Ibid., 8. The study referred to is de Jong and von Hippel, *User Innovation in SMEs*.

127. OECD, *The OECD Innovation Strategy*, 39.

consists of maximizing benefit. As we saw in previous chapters, for the authors of the *Oslo Manual*, innovation is not an end in itself but a means of augmenting the competitiveness of a company. An innovative user, however, can be guided by very different objectives and values when starting out on an innovative process: for example, innovating may be enjoyable, or sharing knowledge with other users (thereby advancing knowledge in general) may be satisfying. These values are not those of the *Oslo Manual*, and this is why the OECD is so reluctant to accept sincerely that users can also innovate, whether they are companies or not. Whenever you believe in a paradigm, you end up developing irrational behavior: following publication of the empirical data in the aforementioned 2009 thematic paper by Innometrics, the OECD and Eurostat should reconsider in a major way their assumed theories because they are being called into question by published data, by user innovation, and, more generally, by the different forms of hidden innovation and non-R&D-based innovation that I have discussed in this chapter.

The Challenge of Measuring Social Innovation

On the basis of the arguments in this and previous chapters, I would conclude that the innovation stemming from R&D is just a part, albeit a very important one, of innovation in general. The linear model is, then, insufficient for innovation studies and policies, as almost everyone now recognizes, although in fact this same model continues in large measure to inspire government and public administration innovation policies. There are multiple forms of hidden innovation that go undetected by the dominant paradigm as represented by the *Oslo Manual* and its annexes. Very important economic sectors such as the CCS, which has a strong social component wherever it is, generate innovation without resorting to R&D. Moreover, the public sector can also be innovative, thereby calling into question the Schumpeterian principle that only companies innovate. As regards user innovation, it offers a real alternative to the *Oslo Manual* paradigm, as we saw above.

Given this situation, there are at least two possible strategies. The first is that favored by the OECD, Eurostat, and the EU: to not renounce the current paradigm and adapt it as far as possible to a changing reality in which forms of innovation emerge that have never been considered as potential areas of study. From Kuhn, we know that whenever a paradigm enters into a crisis, its proponents try to explain as best they can its multiple anomalies, and in order to do so work out ad hoc explanations. This is what happened during the first decade of the twenty-first century. The systemic model was proposed as a replacement for the linear model, but this did not impede the appearance of multiple forms of hidden innovation. This was due to the fact that, at root, the systemic model just evolved out of the linear model, as I have argued on multiple occasions here.

I believe, then, that it is time to opt for a second strategy: to construct a new paradigm from which to interpret and study the diverse forms of innovation. In my opinion, this paradigm is actually being prefigured in current studies on innovation although there is still some way to go before it takes shape, as we saw in chapter 3.

In order to move forward in this direction and even speak about an alternative paradigm, we need to confront an important challenge: how to measure social innovation. Here, then, I will offer some ideas in this regard with the aim of providing an initial sketch for some new instruments of analysis, both qualitative and quantitative, which might allow one to measure different social innovation processes, including user innovation, innovation in the public sector, innovation in the CSS, innovation that originates in civil society, and, of course, innovation promoted by not just profit-making firms but also companies that incorporate social values into their strategies and encourage social entrepreneurship. Given the broad and diverse nature of the sectors under consideration, my proposal here must be general in scope. It is not a question of expressing any disdain for the work carried out by the OECD, Eurostat, and multiple defenders of the *Oslo Manual* paradigm but quite the contrary. I hope to have demonstrated that this paradigm is insufficient when it comes to collecting all the potential diverse forms of innovation, and not even the most significant ones, on the basis of their economic and social effects. I understand that the problem lies in the principles that underlie this paradigm, which I discussed in chapter 2. Given the fact that in this chapter it has been evident that those principles are not just insufficient but ineffective, I believe the time has come to change strategy; instead of patching up the *Oslo Manual*, then, whose 2005 edition accepts the systemic model, we should start from another conceptual model and begin the task of drawing up a new questionnaire and devising new indicators. The 2006 Aho report proposes innovating innovation.[128] For my part, I seek to innovate within studies of innovation and carry out a disruptive innovation. The considerations and proposals on the basis of which I would confront this challenge to measure social innovation are the following:

1. To date, it has not been possible to implement systematic policies to promote social innovation. The United Nations, European Commission, and the White House Office of Social Innovation and Civic Participation have focused on identifying and encouraging good practice in social innovation in specific cases.[129] However, none of these institutions has ever put forward a set of tools by which to measure and analyze social innovation processes. It is true that such processes are very complex and very diverse, and this is why social innovation is such a challenge for innovation studies and policies.
2. To the extent that, within its Innovation Union 2020 strategy, the EU is beginning to promote policies of fomenting social innovation, the need arises to measure social innovation with the aim of shaping these policies. It is the same at the national, regional, and local level. Different innovation systems tend to develop social innovation, with the resulting necessity for data in this regard, both for public policies and for corporate and organizational strategies.

128. Aho, *Report of the Independent Expert Group on R+D and Innovation*.
129. For example, in the form of Social Innovation Fairs organized by UNCTAD (the United Nations Conference on Trade and Development) from 2005 on, the EC European Initiative for Social Innovation (2010), and the Office of Social Innovation and Civic Participation's Social Innovation Fund (2009).

Likewise, policies for fomenting social innovation should be specified, distinguishing them from those that promote innovation in general. And this distinction must be epistemological, methodological, and axiological, as we will see in the following chapter, which concerns the philosophy of innovation.

3. In order to obtain and process those data, it is necessary to conceptualize (define), make operative, and measure social innovation processes. It is not enough to just identify and reward good practices, as has been the case to date. Operational definitions, data, indicators, and, where relevant, indicator systems are required. Yet it so happens that the simple act of enunciating this challenge is problematic because there is no generally agreed definition of social innovation and because of the existence of a diverse number of heterogeneous areas and activities that are classified as examples of social innovation. There is also the problem of the scale at which social innovation processes are produced (and measured). It would not be surprising to find out that measurement instruments differ according to diverse (micro, meso, or macro) scales. Measuring social innovation implies developing a panoply of measurement instruments rather than just one single tool.
4. In order to confront the challenge of measuring social innovation, a flexible conceptual framework is required that is capable of being applied to different types, scales, agents, and processes of social innovation. Furthermore, this conceptual framework must be capable of evolving in line with how social evolution might evolve. Social innovation processes are:
   - Multidimensional, which is why it is necessary to select the dimension that will be measured (as is the case in other indicator systems).
   - Multi-agent (including several stakeholders), which is why they are more difficult to analyze than the *Oslo Manual*'s TPP innovation concept, in which there is just one innovative agent, companies. This particu-

lar feature of innovation processes has been accepted by the systemic model, which is why we must always bear in mind the contributions of this model's theoreticians, especially those who were truly influential when it came to drawing up innovation policies (Bengt-Åke Lundvall, Richard R.Nelson, Silvio O. Functowicz, Geoff Mulgan, Anthony Arundel, Hugo Hollanders, Agnès Hubert, and so on). As I stated at the outset in the prologue, my main area of study is the EU, the body that has designed systematic programs to promote social innovation.

- The model to be followed must be pluralist, and not just because there are various types of innovation, different socioeconomic contexts, and diverse scales but also because there is a plurality of potentially innovative agents, especially in the case of social innovation. Social innovation can be carried out by companies and by other agents. It should even be stated that civil society (the third sector) is a source of innovation in very significant cases (according to both the aforementioned Goldenberg report in Canada and the Young Foundation in the United Kingdom). This leads to the methodological problem of working out who represents that third sector and, more generally, the challenge of surveying several agents in a social value network, rather than just one, as the *Oslo Manual* (and more recently the *Copenhagen Manual*) does.[130] For this reason, a multi-stakeholder methodology is required.
- There is an important consequence of the previous idea: with diverse agents intervening in (social) innovation processes, and with these same agents embracing conflicting or at least different values, such social innovation processes do not generate just one form of value – such as economic value for example – but are in fact axiologically pluralist owing to the fact that di-

130. See Bloch, *Measuring Innovation in the Nordic Countries.*

verse stakeholders maintain their own specific value systems. The key to the plural model I propose here lies in the possibility of introducing *evaluation matrixes* in order to evaluate social innovation processes from an axiologically pluralist perspective. I will return to this point in the final chapter.

- It is not just innovation agents that need to be identified but also those who present difficulties for or obstacles to innovation. These agents also form an important part of value networks. Such agents always exist because of the very plurality of agents involved in innovation processes and the also because of the important fact that any innovation benefits some and harms others, as Schumpeter points out when speaking about creative destruction. Any (social) innovation policies that eliminate or neutralize such barriers or opposition are possibly more effective than those that encourage, emphasize, or publicize (socially) innovative agents. It is not just a question, then, of promoting good practices but also of combating bad practices, which also exist.
- As regards both positive and negative outcomes, a case study methodology should be used in which attention should be paid to both successful and unsuccessful (social) innovation processes. One learns more from failure than from success. The culture of innovation implies specific learning processes, including some that teach how to lose and how to fail, examining the factors involved in and reasons for such failure.

5. Without overlooking my critique of the dominant paradigm and its associated methodologies, the truth is that, in order to construct an alternative paradigm, one must bear in mind those same methods and instruments, precisely because they have, in large measure, failed in that they neither detect nor analyze innovation processes in very important sectors. Innovating in studies of innovation implies learning from the (relative) failures that the method-

ologies of the *Oslo Manual* paradigm have experienced in recent years. The most canonical of these are:

- The *Oslo Manual* for business innovation (the initial TPP model).
- *The Copenhagen Manual* for innovation in the public sector.
- New OECD proposals to measure other forms of innovation, especially those not based on R&D.[131] I already remarked above that in its publication, *The OECD Innovation Strategy* (2010), the OECD states explicitly: "The OECD Innovation Strategy also delivers a new set of indicators, clearly showing that it goes well beyond R&D."[132]
- Nesta's proposals in the same sense.[133]
- The methodologies used by agencies and institutions created by the European Commission (Pro Innova, Innometrics, the Innobarometer, MERIT, and so on), both at a national level (the European Innovation Scoreboard or EIS, subsequently renamed the Innovation Union Scoreboard or IUS) and a regional level (the Regional Innovation Scoreboard or RIS).
- Methodologies used by other countries and certain international institutions to analyze social innovation: the United Nations/UNCTAD, Canada, the United States, Australia, New Zealand, and so on.

6. One should not just focus on the major but also medium-level and small innovations. Social innovation studies must be based on valid tools for social macrocosms, mesocosms, and microcosms. In general, these tools will differ from one another, but also be closely related.

Seventh, there is no social or technological innovation without

131. OECD and FORA, *New Nature of Innovation*; OECD, *Measuring Innovation*; OECD, *The OECD Innovation Strategy*.
132. OECD, *The OECD Innovation Strategy*, 4.
133. For example, NESTA, Beyond the Creative Industries, and Creating Innovation; Hughes, Moore, and Kataria, *Innovation in Public Sector Organisations*; and Flowers et al., *Measuring User Innovation in the UK*.

a certain degree of social acceptance for such innovation. Therefore, one should not just speak about the impact or social diffusion of innovation but also, and above all, about its acceptance (or rejection) on the part of whatever group or institution. Innovation processes are socially complex and one must examine the different facets – both positive and negative – of these processes.

I could add more points, but some of them would require prior reasoning, which will be forthcoming in the next chapter. These initial ideas provide an appropriate start from which to confront the challenge of conceptualizing, identifying, evaluating, analyzing, and measuring social innovation processes.

5
TOWARD A PHILOSOPHY OF INNOVATION

Defining Innovation?

This chapter addresses innovation from a philosophical perspective. Both the philosophy of science and that of technology are well developed disciplines at an international level. And the same is true for the history of science and technology. Yet there is no discipline of the philosophy of innovation or any history of innovation, even though some books are just starting to appear on the subjects.[1] Moreover, if we compare studies of social innovation with those of science, technology, and society (STS), the latter bodies of work are much more developed while the former has just appeared, in a similar way to social innovation policies.

Many people would imagine that one should first ask what innovation is in order to develop a philosophy of innovation. I will indeed pose this question, but I will actually start with another question: is it possible to define innovation? Obviously the answer is yes because there are many – in fact a great deal of – definitions

1. See for example Johnson, *Where Good Ideas Come From* and (in Spanish) Maidagán et al., *Filosofía de la innovación*. As regards the history of innovation, the most systematic approach is that of Benôit Godin's various publications as part of the project on the intellectual history of innovation at the INRS (Institut National de la Recherche Scientifique, national institute for scientific research) university in Montreal, Canada. As of 2012, he had published a dozen research papers on the topic. See for example his *Social Innovation*.

of innovation.[2] However, given the state of the situation, there are grounds for doubting whether it is actually possible to have one single and minimally acceptable definition, for various reasons:

First, the field of innovation studies is yet to be delimited, and continues to grow both in terms of extension and diversity. I will argue in this chapter that natural innovation, in which humans play no part, also exists. Any philosophy of innovation must take this kind of innovation into account, yet almost none of the available definitions contemplate this area of study. In extending innovation studies to non-human worlds any research is made more complicated (initially at least), but in doing so one encounters an original conceptual framework.

Second, innovation processes are complex entities whose structure is by no means easy to analyze. There are those who believe that inventing and innovating are one and the same, and others who practically identify creativity with innovation. Some of us conceive of innovations as processes, while others see innovation as just the outcome of such processes, for example in the case of new artifacts that derive from a technological innovation process.

Third, something is new only in relation to what existed beforehand, so innovation is relative to a previous state of things, which in turn makes it quite difficult, philosophically speaking, to define what is actually new.

Fourth, there are various dimensions to anything new, as pointed out by Jon-Arild Johannessen, Bjørn Olsen, and G.T. Lumpkin: "To define and measure innovation better, we investigated three dimensions of newness: what is new, how new, and new to whom?"[3] Some things are new for some people but not for others, or new in a specific place or time but perfectly well known in other places and other times.

2. The accompanying text for the "innovation" entry in Wikipedia states, somewhat ironically: "the number of definitions of innovation seems to grow even faster than the number of researchers and practitioners in innovation. What is more, the term innovation seems to refer to fundamentally different matters sometimes, which is the case if one colleague calls a new product an innovation while the other two consider an innovation either the process that leads to the new product or the products market diffusion." Accessed August 17, 2012, URL: esw.wikipedia.org/wiki/innovación.

3. Johannessen, Olsen, and Lumpkin, "Innovation as Newness," 20.

For all these reasons, I will initially remain skeptical about the notion of seeking out the "true definition of innovation." We do have at our disposal, though, several definitions of innovation that are to varying degrees sufficient, useful, or operative. Some are valid for some aims and subjects, and others for others. As a whole, that diversity of definitions will allow me to elucidate the concept of innovation. But there is no one single universal definition of innovation. Let us see why not:

1. No type of Platonism will be of any help when it comes to defining the concept of innovation. Innovation is not an idea but a process that implies agents and actions as well as specific contexts and settings. Plato's *kosmos noêtos* (intelligible world) and the Christian Kingdom of Heaven are unchanging.[4] There is no movement, alteration, or innovation there. Innovation processes always take place in specific spatial and temporal situations. These processes may originate in some innovative idea, but as Schumpeter says, one should not confuse invention with innovation. In chapter 3, I critiqued Geoff Mulgan's definition of innovation as "new ideas that work," originally drawn up for the Young Foundation and later adopted by the British government in its document *Innovation Nation* (2008). New ideas may be a necessary condition for innovation, but this is condition is not sufficient in itself. Moreover, there are natural innovations that do not depend on humans, and as such there is no accompanying idea. Essentialism and idealism are not suitable philosophical frameworks for reflecting on innovation.
2. Furthermore, one must bear in mind that concepts evolve over time and meanings change, especially when they are transformed into multipurpose words that are valid for expressing different things, as in the case of the word "innovation." I previously demonstrated how the technical concept of innovation has changed in recent decades and now even includes the notion of social innovation. Any philos-

4. I am referring to the Kingdom of Heaven as conceived by Augustine of Hippo (St. Augustine). See his *City of God* in this regard.

ophy of innovation must be based on an evolutionary epistemology that accepts changes in and the evolution of concepts according to how they are used by different people, institutions, and organizations; this is a strong reason for mistrusting any notion of an "authentic definition" of the concept of innovation. Put another way, a concept is not a static idea. It does not always mean the same thing. Elucidating a concept of innovation is a philosophically complex task that has little to do with defining it by means of a combination of words that would pin down "the essence of the concept." As Gottlob Frege demonstrates, concepts are functions that change meaning according to how they are applied by different semantic domains.[5] Thus, as a concept, innovation is plural. There is no one innovation, but many innovations. And this word, like so many others, has a number of different meanings. There are not one but several ideas of innovation and several definitions of innovation – and not just in dictionaries but also among those who study innovation. I will call this semantic diversity or, if one prefers, *epistemic pluralism*. This does not imply that a word means anything and is valid to express anything. Quite the contrary. In contrast to semantic monism (there is one, and only one, idea for every concept), epistemological pluralism affirms plurality at the very core of the logic of concepts. I will not go too deeply into these questions, but I want to make it clear that the philosophy of innovation embarked on here is pluralist in outlook, both in regard to its epistemology and in terms of its semantics. And this is the basis on which I propose a pluralist model for innovation studies.

3. My recent comments hint at another philosophically important question: we are users of a language and this is a social and distributed entity. Therefore user innovations in every language are possible, although few of them have their own assigned name for the author of the innovation in question. These terminological innovations could be

5. Frege, "Function and Concept."

considered like *small social innovations* in the event of them being accepted and incorporated into common speech. There is, then, a case study to be done that has never appeared among those in social innovation: the emergence of neologisms.[6] They are of undoubted philosophical interest and not just because of all this happens in the realm of language — one of the principal cultural forms of any society — but also because they are the result of a multi-phase process: from the invention of the neologism in question (a new idea) to its ultimate integration into the accepted lexicology of a language, by way of its progressive acceptance and use by other people and groups. Innovation studies have a new case study in the appearance of neologisms in a language and their acceptance (or not as the case may be) on the part of the speakers of that language. Any speaker of a language can create neologisms and even invent an idiolect. Such language games are quite typical among young people and, more generally, in codified languages, some of which are very technologically significant. However, not all new words, or indeed new codes, are inventions. They must be accepted and used by the other speakers. In other words, on the one hand they must necessarily be diffused in such a way that the word in question is known and accepted, and on the other, *they must be used by other people*, not necessarily everyone but certainly a sufficiently wide section of the population. It goes without saying that these small social innovations have no economic value, but they do still generate value: at the very least, a value of use, given that they are used widely by speakers of a language because they are useful when it comes to designing, understanding, communicating, recording, and any other function of a language. It is worth mentioning this example of a small social innovation, neologisms, in order to illustrate some of the singularities of the concept of innovation and how it differs in regard to inventions and new ideas.

6. The disappearance of other words, which normally occurs as a result of lack of use, could also be a topic of study.

4. I would draw one conclusion from this brief reflection: innovation should not be confused with the new. For something to be an innovation there are other requisites besides novelty. I will return to this difference later.
5. What I have just observed as regards terminological innovations is also valid for the construction of new meanings and definitions. A neologism (a completely new word in a language) is not the same as a new acceptation, a new inflection, or a new combination of previously existing syntagmas in a language. The same could be said for definitions, an area in which one can also innovate. Contributing a new definition, if this achieves a certain degree of acceptance, implies a *conceptual innovation*. However, not all new definitions are conceptual innovations but far from it. In order to be a conceptual innovation, a definition must be accepted by other people after evaluating its suitability, usefulness, functionality, or any other kind of evaluative criterion that may be relevant when it comes to comparing different definitions of the same concept or term. This is one of the main reasons to state that there is no "true" or "authentic" definition of innovation. Instead, there are more or less widely used and accepted definitions. New definitions of innovation can always be suggested. But the problem lies in the fact that those new definitions must make sense and, thereafter, be used by people who are interested in innovation issues.
6. These syntactic, semantic, and pragmatic problems offer a means of entering into a philosophy of innovation through the philosophy of language. Whoever proposes a new definition of innovation has not carried out any innovation yet. Conceptual innovations are complex processes in which different agents take part: one should not confuse them with having new ideas, being intuitive, or being creative. Conceptual innovations are of great philosophical interest, due to the fact that one of the contributions philosophers have made throughout history is proposing new concepts, some of which have enjoyed wide social accep-

tance to the point of being widely embraced by a society (the French Revolution and the *philosophes*, for example) or even internationally. In sum, when thinking about definitions we come across an important question for the philosophy of innovation.

7. There are more arguments to counter the notion that a definition of innovation exists, understood in the sense of a definition par excellence, the "good" definition. Any human being can propose a new one, regardless of whether this proposal will be successful or not. In the art of defining, the difficult thing is not proposing something new but innovating, making others accept what is being proposed. Some definitions are better or worse than others according to the criteria of evaluation we might use. One such criterion is novelty or originality, but there are others that are equally or even more important: scope (in relation to what is being defined), accuracy, simplicity, consistency, fruitfulness, and so forth. I mention these evaluation criteria because they are values Thomas S. Kuhn considered typical of science.[7] Philosophers like Larry Laudan and Hilary Putnam term them *epistemic values* or *cognitive values* given that they permit the privileging of one kind of knowledge over another on the basis of being more accurate, rigorous, general, simple, productive, or suitable for what is under investigation.[8] To be sure, Kuhn never included the truth in his list of the permanent values of science because he was adamant, as a critical reader of Popper, that scientific knowledge with an empirical base is always provisional and that its principles can be modified. And I would say the same thing when it comes to definitions, in this case, definitions of "innovation." The question of whether something is an innovation or not cannot just be explained empirically, as the methodologies of innovation studies demonstrate, from Schumpeter to von Hippel by way of the *Oslo Manual*. In short, there is no "true" definition of

7. See Kuhn, *The Essential Tension*.

8. See L. Laudan, *Science and Values* and H. Putnam, The Collapse of the Fact/Value Dichotomy.

innovation. Among the many that do exist, and any new ones that may be proposed, some will be more productive, accurate, general, rigorous, suitable, or simple than others. And we will choose one definition over another according to these epistemic criteria of evaluation. That will depend, in turn, on what we are doing. If we want to promote a specific innovation policy, we will define that concept in such a way that it coincides with our political objectives and actions. If we are economists and carry out studies of the economy of knowledge, it is highly likely we will prefer a definition like that of the *Oslo Manual*, which mostly pays attention to the impact of innovations in markets. Meanwhile, sociologists of innovation would possibly prefer definitions that include social aspects of innovation processes, and so on. One could conclude then that, to be sure, there is and there will be a multiplicity of definitions of innovation. By the way, I would also conclude that the act of defining itself is guided, for example, by the aforementioned epistemic values.

8. All this leads me to infer, likewise, that the definitions of innovation that we consider or use will be *charged with value*: at the very least, epistemic values. The epistemological pluralism of which I spoke earlier is based on axiological pluralism, which starts from three major hypotheses: that human actions are guided by values; that there are different types of values; and that there are values specific to science, or epistemic values. These are the axiological reasons why I contend that there is no one single definition of innovation but instead several, and that some are preferable to others according to the evaluative criteria we use. One of the central themes of the philosophy of innovation is that of the axiology of innovation, as we will see later.

I will conclude this section by making a provisional observation in this regard: What are the values that might lead us to prefer one definition from a philosophical perspective? And also, does a definition of innovation exist that might be considered philosophically preferable to others?

In order to answer these questions, I should first clarify (however briefly) what philosophy is. I understand philosophy in the etymological sense of the term, as the desire to know. That said, there are many ways of knowing and forms of knowledge and, at least in principle, they are all of interest to any philosopher. Given that different disciplines cultivate specific fields of knowledge (physics, biology, engineering, sociology, economics, and so on), philosophy must inquire into all the knowledge they generate, converting it into the philosophy of science, technology, or the social sciences. And because the natural languages are also deposits of knowledge, a philosophy of language must also be carried out by paying attention to the knowledge encapsulated in each language (although repudiating its errors, which also exist: not everything is so clear in languages). The philosophy I am interested in is seized with admiration for knowledge and understanding and examines them, but such knowledge and understanding is possessed by others, not philosophers. Philosophers do not know, nor must they know; it is enough for them to want to know. Any philosopher who believes he or she knows something substantive is no longer a philosopher and has become a pedant or an erudite, or both things at the same time. Philosophy is inquiry, the desire to know.

This way of understanding philosophy, which for my part I advocate, is typically termed adjective philosophy or the "philosophy of (something)" in contrast to substantive philosophy (metaphysics). There is a pejorative connotation to this distinction because some academics think that substantive philosophy is authentic and the field concerned with "true philosophical problems." I consider the scorn of those academic philosophers to be praise. I am pleased to practice adjective philosophies, of which there are several, in this case, the philosophy of innovation.

The philosophy of innovation starts from the basis of the knowledge and understanding that others have developed, whether because they have been innovators or because they have carried out innovation studies. From the former point of view, one should take an interest in innovative practices or, in other words, the set of actions (and agents) that propose innovations, whether they are successful or not. From the latter point of view, it is a question of considering studies of innovation, which have proliferated in recent decades, as we have seen in the preceding chapters. On the one hand there is a philosophy of innovative practices; on the other there is a philosophy based on studies of innovation that have been carried out by economists, sociologists, political scientists, historians, management experts, or any other people that have taken an interest in innovation and thought about it, contributing some understanding in this regard. Note that I use the term "knowledge" and not "opinion." Anyone can have an opinion on innovation and a criterion in this regard, but that is not the same as possessing knowledge about innovation. One thing is *episteme*; another is *doxa*. Moreover, anyone can be creative and have many new ideas, but very few people manage to be innovative. As noted already, at the outset of his work, *Open Innovation*, Chesbrough states plainly, "Most innovations fail."[9] Research is necessary into which forms of knowledge, abilities, and skills characterize innovative agents, whether these are people, groups, companies, organizations, institutions, or other kinds of bodies. There should also be some inquiry into the obstacles to and difficulties for innovation, which has not been examined so thoroughly. Because philosophy is the desire to know, the philosophy of innovative practices will look into which kinds of knowledge and understanding are at stake throughout innovative processes. In my case, I will focus especially on the values that intervene in these processes, given that I will elaborate a philosophy of innovation from an axiological perspective. As regards studies of innovation, I will pay particular attention to the epistemic values that guide these studies.

I understand philosophy as transversal research whose object lies in the knowledge and understanding generated by other occupations or professions. Given its object of study, the philosophy of innovation must also take an interest in innovative practices, regardless of where they originate. In this event, there emerges a *practical philosophy of innovation* instead of just second-rate philosophical studies or the previously mentioned "philosophy of."

9. Chesbrough, *Open Innovation*, xvii.

Three Innovators of Philosophy: Aristotle, Bacon, and Leibniz

Philosophy addresses concepts, especially if they cross languages and have been used in several cultures and at different times. Aristotle coined some important ones, taking them from the Greek language of his era and imbuing them with precise technical meanings. In other words, he defined them rigorously and transformed them into philosophical concepts. Down the centuries, some of these Aristotelian concepts have permeated Western languages through terms such as "categories," "genres," "species," "predicates," "subjects," "attributes," "syllogisms," "deduction," "induction," and so forth. Aristotle was a great conceptual innovator because his contributions enjoyed wide social acceptance in Ancient Greece, Ancient Rome, and Medieval Europe, leaving his mark on philosophical and scientific thought for centuries.[10] They were sustainable innovations used by many languages and cultures over time. Nowadays, most conceptual innovations are ephemeral and at times pure fad, at least in the realm of philosophical and sociological thought.

The analysis of concepts has been an important part of philosophical research throughout the history of philosophy. Generating new and productive concepts brings with it epistemic innovations that allow one to see the world from a different perspective. Insofar as many humans appropriate these concepts, use them, and incorporate them into common speech, cognitive and social innovations are produced that shape conceptions of the world. Some philosophical and scientific concepts have changed our way of seeing the world, calling into question previous weltanschauun-

10. The same could obviously be said for other Greek philosophers but in the case of Aristotle, much of his terminology has entered into the grammar of many languages and this is why his conceptual contributions are so important. In particular, the fact that he authored the first great treatise on logic (the *Organon*), together with the adoption of this work as a standard textbook in medieval schools and its use as such for centuries thereafter, were important steps in conceptual innovation processes that originated in Aristotle's work. The same thing happened in the cases of Euclid, Hippocrates, and other Greek thinkers. Conceptual innovation is not an issue exclusive to philosophers.

gen and even destroying or at least marginalizing some of them. The history of philosophy has generated multiple disruptive conceptual innovations. It goes without saying that concepts are not patented, nor do they have a market value, but they are instead part of the common good, which does not prevent those conceptual innovations being significant for humans. In contemporary jargon it is said that philosophy has contributed a *conceptual capital* that has benefited many people, thereby becoming *social capital*.

In 1620, the Lord Chancellor of England, Francis Bacon, wrote the *Novum Organum*, a work of logic and methodology that went beyond the Aristotelian framework. This work is at the root of modern science because it prioritizes observation and the experimental method as well as underscores the importance of techniques, which had been rather undervalued by the Ancient Greeks. For Bacon, technical instruments are indispensable for scientific research and for a better understanding of nature, beyond what our own limited perceptive capacities offer. He also criticizes false concepts and the idols of the cave, the market place, the tribe, and the theater – communal spaces that contribute not knowledge but falsehoods and confusion. Instead of accepting them credulously, philosophy critiques them following the example of Socrates, whose maieutic techniques were aimed at refuting what someone thought they knew until they ultimately reached the point of being conscious of the fact that, in reality, they knew nothing – the moment at which that person's desire to know with all their might could shoot forth. Bacon was not Socratic but he was a falsifier of unjustified beliefs. For that reason he was a great process innovator or *methodological innovator*, which is different from a conceptual innovator.

Bacon was one of the great inspirations behind the modern scientific revolution, culminating later in the work of Galileo, Descartes, Huygens, Newton, Leibniz, and many others. At that time, the conceptual innovations of philosophers left their hallmark on the most diverse sciences. In current terms we would say that Bacon promoted several *disruptive process (methodological) innovations*, as Aristotle did before him. In Bacon's case, those methodological innovations were diffused and taken on board by nascent scientific societies, although they were rejected by many universities and religious sectors (barriers to innovation). Although Bacon was not an especially precise or rigorous thinker, his methodological proposals were very productive in that they were accepted, applied, and reinterpreted by the most prestigious scientists and philosophers.

However, Gottfried Wilhelm Leibniz also stands out as just as much of an innovator. He invented combinatorics, binary language, differential calculus, and dynamics, all of them important process innovations. He was also an innovator of technological products (the mechanical calculator, the wind machine, and so on), organizational products (the decimal system for classifying books, the Academy of Sciences in Berlin, and so forth), and even communication products (his grand *characteristica universalis* project, of which binary language was a part, which is today termed digitalization). Some of his inventions were of great scientific and social use. Above all, this Hanoverian philosopher and mathematician did not just restrict himself to producing all kinds of innovations (whether conceptual, scientific, technical, religious, legal, or institutional); he also advocated a method for inventing, the *Ars Inveniendi*. Attempting to go beyond Aristotle, Bacon, and other thinkers (especially Descartes), Leibniz formulated his *logic of invention* and applied it to producing new concepts. Some of those conceptual innovations failed, while others were successful. He demonstrated a will to innovate at all times, moreover posing the issue of whether or not there was a method to do so.

According to Leibniz, there is not just a logic of scientific discovery but also a methodology of invention. As a good rationalist, he thought deeply about *Ars Inveniendi* (what I would today term *Ars Innovandi*) and advocated a set of rules to encourage invention in different fields of understanding. His proposals mostly leaned toward conceptual innovation and scientific discovery, but he also concerned himself with technical inventions, which he gave great importance to for human and social progress. His objective was very ambitious. It was not just a question of inventing but of doing so rationally, that is to say, in accordance with a method. If we reinterpret his theses today, adapting them to twentieth-century technoscientific research, we find a good source of inspiration to innovate because his target was to arrive at *knowing how to invent*, following a method of analysis and synthesis to do so. Correcting and improving the rules for guiding the spirit of Descartes – centered on the certainty, order, and the correctness of reasoning – Leibniz affirms the inventive productiveness and utility of rational procedure, in other words, the capacity of reason to innovate in all areas. Rational methods serve to explain nature by means of principles and laws and to deductively predict phenomena that necessarily must occur (as well as others that will never occur) according to the *laws of nature*. Yet moreover, rationality must be useful in order to invent, understanding by this term first the generation of new concepts, new principles, and new laws; and second, as a means of thinking up and producing new instruments and apparatuses that might be useful for life in general and that might solve problems, both of a technical and a social nature. In short, I would contend that Leibniz was a great philosopher of invention, not just for his contributions to science, technology, and philosophy (monads, the best of all possible worlds, pre-established harmony, little perceptions, and so on) but also and above all because of his audacity when it came to creating logic and a methodology of invention. Let me evoke briefly what the rules of the Leibnizian method of invention were.

Throughout his life, Leibniz concerned himself on more than one occasion with the *Ars Inveniendi*, ultimately formulating rules to develop it. I will just discuss one of his texts on the subject: "De la sagesse (On wisdom)" (1676)[11]. Here, he proposes rules for judgment, for invention, and for memory. This text is part of Leibniz's interpretation of Descartes's *Regulae* and demonstrates the differences between the Cartesian and Leibnizian methods. According to Leibniz, invention relies on the method of analysis and synthesis, following these rules:

1. One should know what is being researched in depth, delimiting its requisites (its definition and specific properties). It is important to point out that, for Leibniz, one can come across several different definitions of the same

11. Leibniz, *Philosophische Schriften*, GP VII, 82-85.

thing. It is thus a question of inquiring into all of them, one by one, and seeking out a more general definition that encompasses all of them.

2. One must repeatedly apply rule 1 to each and every term that might appear in each of the starting definitions.
3. This process is repeated for the new definitions that emerge until one arrives at terms that may appear impossible to analyze or intelligible in themselves. Having arrived at this point via all the branches in our analytic tree, we know the thing proposed as well as possible because we have found its conceptual requisites, or at least most of them. Concepts are analyzed to different levels of depth in this kind of logical-philosophical investigation.
4. The previous analyses must be repeated over and over again, both in order to identify possible errors and to discover means of analysis that were not previously evident. The objective is to take into account all the general requisites of the thing analyzed. If we are capable of giving conceptual reason to the initial definitions by means of those general requisites, we will have perfect knowledge of the thing – something that happens very rarely, as Leibniz himself highlights. Nevertheless, even though the conceptual analysis may be partial and not arrive at its final term, it is still useful for demonstrating and finding new truths, that is, as a method of invention. It is easier to find new truths from intermediate requisites than to analyze a concept or definition to its conclusion. The important thing is that, proceeding like this, new concepts, truths, and proof are discovered as Leibniz himself did throughout his philosophical, scientific, and technical research.
5. This first phase of analysis must begin with the easiest thing, moving from easy to difficult.
6. One must try to not leave anything out in this phase of analysis. For that reason, Leibniz advises following the dichotomies procedure, which allows one to analyze concepts and ideas in a binary way.
7. The final outcome of the analysis must be obtaining more

simple and general requisites, or at least those we might consider simpler given the current state of our knowledge.

8. Beginning with this collection of quasi-simple requisites, the method of synthesis consists of combining them with each other in as many different ways as possible, which will generate some previously existing notions and also give rise to new problems and notions. Insofar as the analysis has gone further, this combinatorial synthesis will potentially be more or less inventive.

The distinctive mark of the good use of *Ars Inveniendi* lies, on the one hand, in finding good definitions throughout the phases of analysis and, on the other, in producing useful inventions in the phase of synthesis. Analysis must be carried out on the basis of proposing new and more perfect definitions than those provided by our predecessors. Synthesis must justify in both a combinatorial and deductive way what is already known, but moreover it must also supply new theorems, new facts, or new artifacts. There is an initial criterion to determine whether a definition is good or bad: that it should be creative – that is, within the definition itself it should demonstrate the possibility of that which is defined. One could say the same for technical instruments: good conceptual analysis must contribute the means to construct new artifacts.

With all due caution, this methodology for the invention of concepts may be reinterpreted and used today as a source of conceptual innovation.[12] In the case of technological innovations, the idea of combining notions translates into integrating components from different artifacts, and the idea of finding the simplest notions becomes determining the elemental components of an apparatus and simplifying them still more. As regards process innovations, this implies an in-depth analysis of the basic components and agents involved or, in other words, delimiting the value chain. Of particular interest is the rule that advocates gathering as many different definitions as possible of the notion being examined before analyzing them and the subsequent combinatorial synthesis. Technical invention would be equivalent to identifying the previ-

12. I am alluding to the work of von Hippel, *The Sources of Innovation* and *Democratizing Innovation*.

ous attempts of other inventors, analyzing them, reducing them to their elementary components, and then combining those components in every possible way until finding the best combination for a given context. For me, those proposals have a lot in common with Schumpeterian recombination. Many of the possible combinations are not usually viable in a given context, and some are even literally impossible; others are possible and even viable. Evaluating that viability implies axiological analysis, in which the difficult thing is to choose the priorities and find the suitable moments for each innovative step. As regards technological innovations, the dominant criteria used to prioritize some over others are utility, functionality, cost, usability, acceptance on the part of users and customers, maintainability, and so on. When the Leibnizian *Ars Inveniendi* was applied to technological, social, and legal issues, to mention just three areas in which it was applied, it was always adapted to the context in which those inventions were proposed. Even so, it was a general method.

I cite these three thinkers because they can be the inspiration for a current philosophy of innovation. In order to draft this philosophy, their thinking would have to be reinterpreted, elucidating the concept of innovation, fomenting innovative experimentation (which differs a great deal from scientific experimentation), and reinterpreting their thought, adapting it to what innovation is at the beginning of the twenty-first century. I will continue, then, to focus on Leibniz with the aim of briefly inquiring into which suggestions from his *Ars Inveniendi* might be valid for a potential *Ars Innovandi*. Innovation studies have a scientific or, perhaps more accurately, technoscientific component, but innovative practices require capacities, abilities, and skills. Because of this, from now on I will talk in terms of the *arts of innovation* and not just the sciences or technosciences of innovation. And the philosophy of innovation must address each and all of them.

In this regard, I will make the following initial suggestions, which I recognize to be fully open to modification or improvement:

1. Just as the conceptual innovations of Leibniz were designed to generate new ideas, his inventive method applied

> to other areas of knowledge must be based on generating value, but according to different kinds of value. Innovating implies increasing the level of satisfying values such as efficiency, social utility, equity, social inclusion, elegance, beauty, applicability, profitability, competitiveness, and so on. Innovative practices have little or nothing to do with the search for truth (except, precisely, in the case of conceptual innovations), but some of them can create value, as we saw in chapter 3.

By means of an example, let us see how Leibnizian rules might be applied to economic values. According to the *Oslo Manual*'s conception, economic and business values – for example, profitability, productivity, and competitiveness – are the most important values associated with innovation. Firms do not innovate for the sheer pleasure of inventing new things or in order to obtain new knowledge, but above all to obtain economic benefit and be more productive and competitive in the market, which is where innovations are implemented according to the *Oslo Manual*.[13] The dominant axiological criteria when it comes to evaluating innovations in the business sector are based on the creation and/or increase of specific economic values (growth, competitiveness, profits, productivity, market share, stock market capitalization, expansion into new markets, and so on). In this case, a philosophy of innovation inspired by Leibnizian thinking would advise companies to apply the rules of *Ars Inveniendi*, translating and adapting them to their own particular activity. This would not consist of analyzing concepts or of generating new truths but of analyzing market situations and their own productive processes with the aim of generating wealth, gathering data in this regard rather than definitions. Interpreted in this semantic and axiological context, Leibniz's first rule would advise an empirical inquiry into the resources (whether financial, human, technological, and so on) a firm has and "defining them," or evaluating and quantifying them. On the basis of rule 3, "other

13. Innovation policies can add other values of a political or social nature to technological innovation processes: for example, generating employment and social well-being, sustainability, social inclusion, and so on. This would be a case of systems of politico-economic values or socioeconomic values.

definitions" of these goods would have to be taken into account such as the resources at the disposal of other competing firms as well as the situation of the markets and social capital in the environment in which the firm operates. Accessing the data and knowledge of rivals is a key operation when it comes to defining innovation strategies. Falling back on "more general definitions" would be equivalent to gathering together and integrating firms and resources until there was a sufficiently sizeable "critical mass" or "market share." This would also imply practicing the kind of open innovation defined by Chesbrough:

> Open innovation is a paradigm that assumes that firms can and should use external ideas as well as internal ideas, and internal and external paths to market, as the firms look to advance their technology. Open innovation combines internal and external ideas into architectures and systems whose requirements are defined by a business model. The business model utilizes both internal and external ideas to create value, while defining internal mechanisms to claim some portion of that value.[14]

The key to the Leibnizian method lies in defining those ideas, which in an economic context is equivalent to expressing them as quantitative data that should be workable and combinable with each other. I am obviously alluding to the methodology based on indicator systems. Those ideas must be combined with a view to innovating and generating value, whether such ideas (data, indicators) come from inside or outside a company, including those of its competitors. Once analyzed, strategies must be devised that generate value (in this case, economic value). The synthesis phase begins with recombining both internal and external data and knowledge, and leads ultimately to the establishment of management and business models that, if carried out successfully, will generate benefits for the company. Potential errors and mistakes must also be analyzed, evaluating the results that derive from this or that strategy. In any event, business innovations differ completely from conceptual innovations. In the latter case, the analysis of definitions and the subsequent combinatorial definition of terms generates new concepts, some of which will be suitable

14. Chesbrough, *Open Innovation*, 1.

and useful and others not. The analysis of new concepts must be syntactic, semantic, and pragmatic. In the case of firms involved in markets, defining new ideas implies expressing them in the form of data, for example marking out objectives and quotas to be achieved in order to increase productivity or market share. A firm defines its innovative ideas when it quantifies them, establishing quotas to be achieved (benchmarking) and designing strategies to increase value, in this case in order to improve the accounts of the results. These are not logical but empirical definitions, based on economic data. One can only verify whether an innovation has been successful or not in the markets, but it is essential to have such data and the expected quantitative results on hand when setting up an innovation process, later comparing them with the real results achieved. Whoever obtains the best results in this field of analysis and empirical synthesis innovates, and this is not logical or conceptual but commercial and entrepreneurial. Substituting epistemic for economic values, the rules advocated by Leibniz still make sense, regardless of whether they have been reinterpreted or reformulated. The Leibnizian method would have as its objective the elaboration of a rational theory of innovation, in which the setting for corroborating the new concepts would not be *noetic*, nor even linguistic (with its dictionary definitions), but instead markets, whether local, regional, national, or international. On the basis of economic data, Leibnizian rationalism would proclaim the slogan "*calculemus!*" which in this sense might be translated as "let's calculate!" Nowadays, it is computers that have the capacity for massive data analysis. That is why we live in the age of the technosciences, as we saw in chapter 1.

The procedures deriving from the Leibnizian *Ars Inveniendi* are not so different from those that many economic agents practice, especially in financial markets. The potential impact of a new financial product in a specific market can be calculated, although in order to do so one must resort to complex mathematical models. The main problem, in effect, is uncertainty. Agents act in uncertain conditions in markets, and for that reason their management models must be based on uncertain situations and scenarios. Rationalist philosophers based their methodologies on certainty, which is at the root of the logical analysis of concepts. While Leibniz's *Ars Inveniendi* may be reinterpretable nowadays, and may even be applied, the philosophy of innovation cannot be based on such simple and mechanical combinational models as those proposed by Leibniz during his lifetime. A philosophy of innovation must pay attention to different types of values (economic, technological, social, legal, political, and so on) and not just to epistemic values. Many of those values can only be increased in situations of uncertainty, without there being any winning strategies a fortiori. Innovation processes are complex, as are the data, the indicators, and the models that simulate them. For that reason they are uncertain.

2. We saw in the previous chapter that the *Oslo Manual* acknowledges that there are innovations not orientated toward the market, such as those produced in the public services: education, health, government, and so forth. In this case, the values to be taken account of are not economic- or business-oriented but of a social and political nature. Quantifying such values is not easy, although at times it is possible to do so: take a political innovation, for example (a new leader or slogan, an uncommon coalition between parties, or a social movement such as the civil rights movement in the United States), whose success or failure might be measured primarily in terms of its capacity to mobilize people, then as regards the voting intentions of these people, and finally in the votes they actually cast at the ballot box. In principle, it is possible to measure the relative force of such an innovation and quantify it in relative terms. Surveys of public opinion and social perception likewise allow one to measure, on an ordinal scale, the level of satisfaction and acceptance on the part of people and users when it comes to evaluating public services or political leaders, to mention just two examples. One could also reinterpret, in this case, the methodological rules of the Leibnizian *Ars Inveniendi*. Research would thus focus on users in the case of service innovation and on citizens and the governed in the case of political innovation, always assuming that the relative strength with which a new political party or new

social movement emerges is measured.

Surveys are the current organons that allow us to "define" (quantify) preferences and the level of satisfaction with one proposed innovation or another. Repeating those surveys periodically, as is typically the case (for example, the Eurobarometers, the public opinion surveys carried out by the European Commission in the EU), is equivalent to applying rule 4 of Leibniz's *Ars Inveniendi*. Of course, one never arrives at perfect knowledge of the thing being inquired into, but one can advance in the analysis of social, political, and administrative phenomena. Clearly, rules 5 (move from the simplest to the most complex) and 6 (proceed by means of dichotomies) are basic for these kinds of inquiries into political and social reality. The combinatorial synthesis of rule 8 could be compared to the methods of aggregating and disaggregating results practiced in empirical sociology. In sum, here as well, although more complicated than in the case of innovations oriented toward the market, the rules proposed by Leibniz may serve as a source of inspiration in order to generate and evaluate social, political, and administrative innovations. Studying the good practices of other agents would be comparable to rule 3. And benchmarking would imply arriving at a point at which one would have to acknowledge that the provisional analysis had finished, given that the current state of knowledge (and innovation processes) cannot advance any further.

Such reflections demonstrate that classic thinkers can serve as a source of inspiration for a philosophy of innovation. Just as Aristotle conceived metaphysics and just as the philosophy of science can be considered a metascientific study (because it is based on scientific and technological practice as it demonstrates),[15] so the philosophy of innovation takes the shape of a "meta-innovative" discipline that must be capable of analyzing the different kinds of innovation that are effectively produced and, moreover, must be capable of contributing a method to integrate those analyses into a conception of innovation that should be applicable to all or at least most of them.

15. See Moulines, *Exploraciones metacientíficas*.

A Naturalized Philosophy of Innovation

The recent boom in social innovation and undoubted link between technological and social innovations has pushed some authors to believe that all innovation is social. In a conference on social innovation organized by the University of the Basque Country in late March 2011, one of the speakers, Jesús Sebastián, stated, "All innovations are social in their processes and many of them in their results."[16] A statement like this very satisfying for sociologists of innovation and even those who study science, technology, and innovation from a perspective of social constructivism, but for the philosophy of innovation I am presenting here it is an insufficient thesis. Innovation processes precede the appearance of human beings because they go back to the emergence of new cosmological entities (galaxies, stars), geological entities (continents, volcanoes, islands), and biological entities (physiological systems, living species, ecosystems, new environmental conditions, and so on). I advocate a *naturalized philosophy of innovation* that does not just address technological and social innovations but also examines (and from the same conceptual framework) evolutionary or natural innovations, which can contribute much knowledge to the study of innovation in general.

As we saw in chapter 2, evolutionary economics is at the core of the innovation studies of Nelson and Lundvall, the proponents of the systemic model. These authors use Lamarckian evolutionary models, but other economists have used and continue to use Darwinian models. For me, I would add a third symbiogenetic evolutionary model, developed and disseminated in recent years by the biologist Lynn Margulis. The recently deceased Margulis always said that the idea of symbiogenesis was first suggested in 1910 by Konstantin Mereschcowsky, a Russian biologist at the University of Kazan who studied the origins of eukaryote cells in bacteria. From the pluralist perspective I advocate here, though, neither the work of Darwin, Lamarck, or Margulis fully covers the evolutionary biology field of study. Indeed, in the late nineteenth century Peter Kropotkin had already maintained that mutual aid

16. Later published as Sebastián, "¿Hay alguna innovación que no sea social?" 48.

is a mechanism that explains many evolutionary phenomena, meaning that the Darwinian struggle for survival is not the only explanatory model for the evolution of species. Instead, it should be implemented together with cooperative models of survival.[17]

When evolutionary theory is translated into economics and the study of innovation, there is a much clearer need for a number of different models. Darwinian models may be of use when it comes to analyzing innovation in certain markets, such as financial markets for example; Lamarckian models are valid for other types of companies and markets, for example those based on R&D or for innovation in the public sector; and symbiogenesis models are most useful when it comes to studying user innovation as well as innovation stemming from cooperative firms and organizations, which are common in the third sector. There are, of course, other mixed models, in innovation studies if not in evolutionary biology. The naturalized philosophy of innovation I present here is pluralist in the sense that it permits the inclusion of various evolutionary models – rather than just one, whether Lamarckian or Darwinian – in order to analyze innovation processes. Whatever the case, here I will now focus on non-social forms of innovation. Before beginning, though, it is worth recalling the basic metaphor at the root of evolutionary economics. Ángel L. Arboníes presents this in the following terms: "I am going to think of companies as biological systems that evolve. Organizations are and behave like living beings, and in order to survive they must coexist with their environment, adapting to and coevolving with the setting."[18]

This metaphor implies a natural dimension to the business world in that it conceives of economic competition as analogous to the struggle for life in the natural world. As Arboníes is one of the current experts on the Mondragon Corporation cooperative, he does not opt for Darwinian models but instead chooses coevolutionary models in which the important thing is not adapting to the environment (as if it were something external to the company) but coevolving with it. It is the case, however, that occasionally there are radical innovations in social environments: one canoni-

17. See Kropotkin, *Mutual Aid, a Factor of Evolution*.
18. Arboníes, *¿Innovación o evolución?* 47.

cal example is the Internet because it generated a whole new social environment — the third environment — in which humans relate to and interact with one another not in person but indirectly and online. In cases like this, a firm (and indeed any agent) must e-migrate to the new social space, first of all adapting to it and then coevolving with it. Moreover, the disappearance of a firm can be thought of as an innovation in the same way that one might think about the birth or death of a person, always understood from an evolutionary perspective. In short, the idea of coevolution in the business world is appealing, although there are processes that imply greater disruption and that must be examined on the basis of other models.

Let me get right to the point. The naturalized philosophy of innovation goes beyond evolutionary economics and social innovation because it also addresses natural innovation, that is, not just innovations generated by human beings or by firms. Innovations that originate in the animal and plant world are of great interest, but there are other simpler living beings, such as bacteria, that are significant for the naturalized philosophy of innovation. Prebiotic chemistry studies the emergence of natural innovations in inert environments. And of course there are other innovations that do not even take place on planet Earth, such as those investigated by astronomers. In sum, besides social innovations one can differentiate many other forms of innovation from a natural perspective: human (individual), animal, plant, those produced within inert material on planet Earth, and those not produced on Earth at all. People have already begun to study some of these innovations and they make up what Steven Johnson terms "the natural history of innovation."[19] The discipline of the history of science dates from the early twentieth century and was soon followed by the development of a related field of study: the history of technology. However, the history of innovation remains a field to be developed.[20] In order to begin this task, the notion of innovation itself (as well as its different classes) needs to specified, at least in minimum terms. I believe the philosophy of innovation can offer a great deal con-

19. See Johnson, *Where Good Ideas Come From*.

20. Benoit Godin is making important contributions to that field. See Bibliography.

cerning this task, and I will therefore make some initial suggestions in this regard.

I will begin with natural innovations, which have been to date the least considered of the different types of innovation. The history of technology and even the history of science serve as a basis for certain innovations – specifically, technological innovations and those based on R&D. However, we have seen throughout the present work that there are other forms of innovation. Studying these from a historical perspective is a pending task, and these studies, when they come, should be linked to philosophical explorations of innovation. Let us see how.

Let us first consider life from the perspective of innovation studies: there is no doubt that, seeing its consequences throughout history, life itself has supposed a disruptive innovation for Earth. There is no scientific consensus about the origins of life, but there are several theories in this regard: from the boiling chimneys or vents of undersea volcanoes (Günter Wächtershäuser), from primordial soup (Stanley Miller), from tidal pools (Darwin), from meteorites originating in outer space (Fred Hoyle), and so on. Nevertheless, current theories of innovation stress the fact that there is no innovation without diffusion. For that reason, it is not enough to just know what the origin of life is in order to examine it as an innovation. Instead, one must also concern oneself with the diffusion of life all over Earth or, put another way, natural evolution. The issue is not just the emergence of life in its most elemental and then more advanced forms. Life has been a disruptive innovation because it has subsisted, reproduced, developed, and diversified, to the point of covering much of the earth (the biosphere). The continual repetition of certain phenomena has been a necessary condition for this to take place: for example, the rising and setting of the sun, with its effect on landscapes and animals; or that of the moon, with its effect on the oceans and tides. Many of the natural regularities that allow for life to be maintained depend on Earth's movements: that of rotation on its axis and its turn toward the sun throughout the year. Whatever the origin of life, it would not have subsisted, reproduced, or evolved without this relatively stable and repetitive environment that has been facilitated by the regularities of the Earth's movements for millions of years. Put another way, the daily routine of dawn and dusk are requisites that facilitate the emergence of many natural innovations. Life would never have been a disruptive innovation had it not been for its diffusion and evolution all over Earth. The great example of natural innovation is the emergence and diffusion of life as well as, later, the emergence and diffusion of different forms of life. In all those cases it is a question of prolonged space-time processes located in one place or another all over the world or in several places at once. For that reason, one could state that *both technological and social innovations as well as natural innovations are processes and are closely connected to the environment in which they are produced.*

According to contemporary biology, innovation stems from chance rather than intentional processes. Moreover, such processes are not even carried out by specific agents, but instead are the result of the interaction of both active and passive multiple factors. One must therefore acknowledge the *systemic condition of innovation processes*. In certain favorable conditions in a system, and after different changes and mutations (most of which result in no outcome whatsoever), an innovation is sometimes produced that leads to a significant change in that system. Normally, natural innovations considerably alter the ecosystems in which they are produced. This is especially the case at the micro scale, although according to how they are diffused, changes are also produced at the meso and macro scales. In the natural environment it makes no sense to ask whether innovations are good or bad. Because they are processes with multiple agents involved therein, there are those who benefit from them and those who are harmed by them.

We tend to think that natural innovation is demonstrated by means of the appearance of new species and, within each species, new individuals. Following the *Oslo Manual*, we could say that this kind of thinking underlies the postulates of product innovation, which for years dominated economic studies of innovation. For my part, I believe that among natural innovations, process, organizational, and communication innovations – to mention the remaining three categories of innovation I acknowledged when speaking about social innovation – are equally or more important.

The philosophy and history of natural innovation must address these four forms of innovation as a minimum. It may be the case that in the natural environment there are other forms of innovation – an issue that will become clearer with more empirical research – but for the moment, one only need inquire whether these four canonical classes of innovation are produced in nature or not. Let us examine briefly if this is indeed the case.

Process innovations of the first order in natural history are, for example, fermentation, photosynthesis, the use of oxygen for breathing, and organic nitrogen fixation. Natural evolution is not just a question of the emergence of new species or the disappearance of others (creative destruction) but also of the appearance of new and vital functions and processes that are typically shared by several species. These processes are usually embodied in the form of organs that develop new functions or also as old functions that operate in a new way. A history of natural innovation based on the notion of process would be very distinct from natural histories inspired by Aristotelian thought and based on genera and species. Instead of the evolution of species, we would be discussing the evolution of functions and organs, many of which are shared by several species.

One could say much the same for a potential natural history based on the emergence of new forms of organizing life (organizational innovation), such as the emergence of social species. And just as marketing innovations have been expanded to the broader concept of communication innovations, so it would also be possible to establish a history of communication innovation that would include studying the appearance of new forms of communication among living things: for example, the exchange of genetic information among bacteria, pheromones in ants, the emergence of the senses of vision or hearing, the beginning of language among humans, and so on. In contrast to the history of natural evolution centered on species and constructed by Lamarck and Darwin, but also complementing it, we would have three other versions of the history of natural innovations: namely, those focused on process, organization, and communication. Although I will limit myself to just pointing out the existence of these other versions rather than discussing them in any great detail, it remains nevertheless quite clear that the interaction among the theories at our disposal and the dominant conceptions of natural evolution can be productive in both directions.

Studies of innovation, then, incorporate a whole new field of study: natural innovations. And the history of biological evolution might likewise incorporate the aforementioned three new perspectives in contrast to relying on the Aristotelian (as well as Linnaean, Lamarckian, and Darwinian) paradigm that organizes the study of living things on the basis of distinctions among genera, species, and subspecies. This is just one of the contributions to be made by the naturalized philosophy of innovation that I am outlining here. There are, though, other and interesting suggestions to be made. Specifically, in the case of natural innovations, almost all of them originate in microcosms and even sometimes in nanocosms.[21] They appear as small innovations and only with the passing of centuries, and their prior diffusion through space, do these natural innovations appear at a larger scale, generating important changes in mecocosms and even for the earth as a whole. The best example of such scaling up in natural innovations are bacteria — those that generated the terrestrial atmosphere that in turn facilitated the emergence of new forms of life based on breathing and the corporeal combustion of oxygen. First there were cyanobacteria, which extracted hydrogen from water in order to feed themselves and freed up oxygen as a residue. These cyanobacteria produced this chemical reaction from time to time, but then came bacteria that produced the same reaction on a continual basis. Oxygen, however, was also a very dangerous residue for many living things that burst into flames on contact with it. Gradually, these life forms disappeared and because the bacteria continued to feed themselves with the hydrogen from water, oxygen was no longer absorbed and started to escape in the form of gas, accumulating in the atmosphere in varying proportions until it reached the current equilibrium. As a result, according to Lynn Margulis and Dorion

21. There are of course exceptions. One typical example is the impact of the meteor that hit Earth and led to a climatic change that resulted in the extinction of dinosaurs as well as many other animals and plants. This was an innovation at the macro scale that reached that of microcosms.

Sagan, there was major atmospheric contamination due to an excess build-up of oxygen, the direct cause of which was bacteria[22]

Through history important changes have taken place in Earth's environment – far greater, indeed, than the climate change that so preoccupies us today. These changes produced the atmosphere as it exists today, with all the destructive and creative consequences that this atmosphere has had for different species and individuals. There are species that adapted well to the current percentage of gasses in the atmosphere and others that did not. In order to survive, those species that adapted well had to alter their organs and change some of their physiological processes. This is a good example of natural innovation, which facilitated the emergence of living beings like humans and mammals, but which also destroyed many other living beings for which the percentage of oxygen was a lethal poison. Given our bodily makeup and respiratory apparatus, we tend to think of oxygen as a good thing, and we consider the bacteria that generated the atmosphere all that time ago, thereby producing the potential conditions for the emergence of current respiratory apparatuses, as beneficial. Yet what was good for mammals and humans was bad for other living species. And this is the nature of natural innovation processes.

One can make another general philosophical conclusion from these examples of innovation because something similar happens with human innovations: *innovations are good for some and bad for others*. The philosophy of innovation must be connected to a theory of good and bad that accepts the empirical condition of good and bad things and develops methods to analyze them. In contrast to those who consider innovation as something solely good, a natural philosophy of innovation underscores the double-sided nature of innovation processes. This is valid for natural innovations but also for technological and social innovations, which give rise to benefits for some and disadvantages for others. One thing is the ideology of innovation, which conceives it as the potential salvation for all ills and difficulties; another very different thing is the naturalized philosophy of innovation, which confronts the problem of creative

22. See Margulis and Sagan, *Microcosmos*, whose contributions I am summarizing here.

destruction in all its crudeness and drama.

Once the atmosphere was created – which was not the work of anyone in particular but that of bacteria – the potential conditions were in place for the emergence of new natural innovations such as photosynthesis.

We thus take one more step toward a natural philosophy of innovation: in general terms, I will speak about value networks because value chains refer to a specific case.[23] Here I am establishing a general theory of innovation that might be valid for different forms of innovation, including natural innovations. And having at one's disposal a general model that is valid for interpreting and analyzing diverse innovation processes, whether natural or artificial, is one of the objectives of the philosophy of innovation that I am presenting here.

One could discuss many other aspects of natural innovations, but because studies of this topic are still in their infancy we will have to wait until more has been done in this regard before systematizing this field of research. My brief commentary thus far, however, shows that a social philosophy of innovation alone is insufficient and that it requires the addition of a naturalized philosophy based on the following guiding principles:

1. Studies of innovation, together with their root epistemology, must be evolutionary. This implies accepting the concept of evolutionary innovation. The philosophy of innovation must be capable of interpreting, from the same conceptual framework, different forms of innovation, from natural innovations to social innovations by way of technological innovations.
2. The philosophy of innovation must be based on an empirical methodology (involving case studies, distinguishing between different phases in an innovation process, the use of instruments to measure results and their consequences, and so on).
3. Definitions are instruments for research that form part of the toolbox of innovation studies. Some of them can be conceptual (combinations of terms), others quantitative

23. See on this Allee, "Value Creating Networks."

> (combinations of data), and some even operational. The general conceptual framework must be valid in order to analyze and interpret different forms of innovation, even though later one may have to construct other definitions and specific indicators for each of the specific types of innovation.

I will now focus on this third point, returning to the problem of defining innovation. I will, specifically, just concern myself here with innovations generated by humans but I will also bear in mind everything I have noted about natural innovations in the preceding pages. In short, innovation processes are innate to living beings, and especially to human beings!

Heuristic Definitions of Innovation

In the first part of this chapter I offered some reasons for considering that there are many different definitions of innovation, with each being more or less valid according to the value system used by the people who formulate or apply a particular definition. I am now going to explore this subject in more depth, introducing the notion of heuristic definitions, or definitions designed for research of whatever variety: economic, sociological, political, business, historical, philosophical, and so forth. Operational definitions like that of the *Oslo Manual* are of this type and are indispensable when it comes to guiding public policies. Those definitions include instruments to work out who innovates and who does not, as well as to differentiate to what extent firms are innovative. In the best-case scenario, they must be capable of measuring innovation, for example by means of indicator systems. In such cases, the operational definition consists, precisely, of an indicator system so that this very definition changes according to how the indicators are altered, as is typical in the *Oslo Manual*, the Innobarometer, the EU's Union Innovation Scoreboard, and other systems for measuring innovation. The precision of such data, their comparability, and the extent to which they can be aggregated are, in these cases, the necessary epistemic requisites so that any definition of innovation and the instruments used to measure it are acceptable. Yet for someone in business, it is enough to have an intuitive idea of what innovation is in their particular economic environment, regardless of the extent to which that idea is rigorous or not, or whether it can be measured precisely. The important thing for this person is that this idea should be proactive, that is, that it act as a spur to encouraging innovative projects of whatever kind. Anyone who is involved in the historical or philosophical study of innovation, however, will have to make use of sufficiently general definitions in order to be able to include the different types of innovation in such definitions. This is my main goal in this chapter.

Nevertheless, among the epistemic values mentioned up to now there is one I regard as especially suitable in order to encourage innovation: fruitfulness. As this is a new form of inquiry that does not even make up part of any academic discipline, innovation studies must be developed in several different directions because the actual concept of innovation has evolved in recent years to embrace new forms of innovation (and will, of course, continue to evolve in the future). Therefore any philosophy of innovation that fits the current state of the art must above all use definitions that, as well as being general, are fruitful, with the goal of promoting research. Another requirement, from a philosophical perspective, is that the definitions used should allow for establishing classifications for diverse types of innovation, while at the same time being common to all of them. At the end of this section I will propose a definition of innovation like this, which to my mind will be sufficiently general and fruitful in order to foment studies of innovation as a whole and philosophical studies in particular. However, I do not mean to suggest that this will be the "authentic" definition of innovation or even that it will be preferable to all others in every case. To conclude this section, then, there will be a heuristic definition that I consider valid as a basis for a general philosophy of innovation. And this definition includes natural innovations as well as human, technological, and social innovations.

Before considering all this, though, it is worth recapping some of the different definitions and characterizations of innovation that we have come across so far. First there is Schumpeter's definition, made on the basis of five cases to define the term "innovation":

> (1) The introduction of a new good – that is one with which

> consumers are not yet familiar – or of a new quality of a good. (2) The introduction of a new method of production, that is one not yet tested by experience in the branch of manufacture concerned, which need by no means be founded upon a discovery scientifically new, and can also exist in a better way of handling a commodity commercially. (3) The opening of a new market, that is a market into which the particular branch of manufacture of the country in question has not previously entered, whether or not this market has existed before. (4) The conquest of a new source of supply of raw materials or half-manufactured goods, again irrespective of whether this source already exists or whether it has first to be created. (5) The carrying out of the new organisation of any industry, like the creation of a monopoly position (for example through trustification) or the breaking up of a monopoly position.[24]

In my discussion of the Schumpeterian definition in chapter 1, I attempted to come up with a more general description and formulated the following: Definition 1: *Innovation is doing something new, or doing something that has already been done, but in a different way*. Later, I proposed another definition that derived from the first: Definition 2: *Innovations are new actions (and ways of doing) that lead to significant transformations in something (let us term it X), whether at the micro-, meso-, or macro-scale*.

These two definitions, the provisional nature of which I underscored, must now be reworked because they are valid only for human innovations and do not encompass the natural innovations I have just described. Furthermore, there is no mention of values in these two definitions. As we saw in chapter 3, the creation of value is one of the singularities of innovation processes. And moreover, another deficiency is that these two definitions just refer to actions and do not include processes in their explanations. Above all, they make a lot of effort to highlight the fact that innovations are actions, with the inference that there are agents that encourage such actions and even promote them intentionally. Indeed, the dominant paradigm in innovation studies fully accepts this assumption, precisely because of its Schumpeterian roots. The *Oslo Manual* repeats time and time again that if it is not Schumpeter's entre-

24. Schumpeter, *The Theory of Economic Development*, 66.

preneur, then it is organizations – and more precisely companies, understood as specific firms – that innovate. It was a similar case for the theorization of social innovation put forward by the Young Foundation in 2006 and 2007: that innovations are promoted by people, groups, or organizations whose ends, like their means, are primarily social. This definition, based on means and ends, was adopted by the European Commission for its Innovation Union 2020 strategy and came to be regarded as a canonical explanation of the term "innovation" in the EU as a whole. Both economists and political scientists are always interested in innovative agents, whether they are firms, public administrations, or organizations in civil society. However, a philosophy of innovation must be capable of including innovations without any subject, which emerge as the result of non-intentional (chance) processes, as argued in the previous section. For all these reasons, then, the two definitions I have offered thus far are insufficient for my purposes although they may be valid for other kinds of (less philosophical) research.

We thus come across a more deeply held assumption in routine studies of innovation: there is no innovation without innovative agents (or subjects). The debate over whether people, firms, organizations, or innovation communities innovate will come later. But it is fully accepted that innovation processes are encouraged and developed by subjects. And it goes without saying that this assumption was also part of the thinking associated with the linear model.

Yet my inquiry into natural innovations above, although necessarily brief and succinct, opens up a whole other series of questions and problems. Now I will take another step forward, arguing that innovations are not just actions but *interactive processes*; that is, they are sets of actions that are developed in time through the interaction of diverse agents and that, as a result, generate something new. Those processes are sometimes propelled by innovative agents but on other occasions not: *there are innovative actions without any human agent to propel them*, that is, *innovations without an innovative agent*. From a philosophical perspective, such actions imply an important conceptual challenge. These forms of innovation clearly surpass the Schumpeterian framework as well as other theorizations. The naturalized philosophy of innovation must offer a much more general conceptual framework, which I will explore in the following paragraphs. Specifically, it requires accepting that there are innovation processes without any innovative agents (or subjects) involved in them at all. Put another way, in effect innovations require actions but these are not always intentional. Innovations also stem from the interaction of multiple agents, but without any of these being considered as the one single "innovator" or, much less, the "creative" agent: there are innovation processes that occur in the biosphere that imply no creativity on the part of any subject. Of course, those innovation processes do not come from new ideas, for the simple reason that in natural innovations nobody has any idea about anything, nor is anyone involved in designing innovations, because such innovations are produced by chance out of the multiple interaction of diverse agents and environmental factors. Nor is there anyone to make an idea out of the end result of the process. These are, to put it one way, *blind innovations*. The field of innovations without any subject is a strictly philosophical area of research that requires more in-depth study.

Before moving on to this, I will critically revisit some of the most typical definitions of innovation. I will leave aside those I have already discussed (by Schumpeter, the *Oslo Manual*, the Young Foundation, Nesta, Stanford University, and so on) and instead focus on definitions offered by Wikipedia. Wikipedia is considered by the Young Foundation, von Hippel, and others as a perfect example of social innovation. Its definitions and commentaries do not necessarily come from scientific or scholarly research, but are instead elaborated thanks to the voluntary collaboration of a set of *expert users* who have come to make up an innovation community. My pluralist conception of innovation must bear in mind the definitions that come from these communities of users[25]. In fact, Wikipedia is a very widely used source of information and knowledge. Given that I demonstrated the significance of user innovation in the previous chapter, I cannot overlook such "user definitions." A theory of "innovation 2.0" must necessarily take into account the

25. See Echeverria, Alonso and Oiarzabal, *Knowledge Communities*.

different people who are interested in innovation studies and not just those who are professionally involved in the field like academics, political scientists, consultants, and researchers.

Wikipedia, then, defines the noun "innovation" as "the act of innovating; the introduction of something new, in customs, rites, etc." and "a change effected by innovating; a change in customs; something new, and contrary to established customs, manners, or rites." And as regards the verb "to innovate," Wiktionary defines it as "to alter, to change into something new," and "to introduce something new to a particular environment; to do something new."[26] For the second meaning of the noun Wiktionary makes no allusion to technological or business innovation. Instead, it focuses on changes in customs or rites, giving the impression that the authors of this definition are much more interested in cultural anthropology than in economics or technology.

The second meaning of the verb "to innovate" also stands out: *to introduce something new to a particular environment; to do something new.* The first part of this definition could be valid as regards natural innovations: the introduction of a new species in a particular ecosystem due, for example, to a migration, is a good example of innovation, whether it is produced by natural actions (the effect of the wind, the movement of animals, and so on) or by human action (invasive species, for instance). That said, the verb "to introduce" – which appears in the *Oslo Manual* in the sense of introducing something new into the market – only refers to the second variant of natural innovation, that of human action.[27] The word "introduction" is occasionally valid when there is no human intervention, but it seems more prudent to speak about the appearance of something new in a particular environment, including its potential emergence, which is not due to anyone except the interac-

26. Wiktionary, accessed August 16, 2012.

27. From this perspective, genetic engineering and nanotechnology would also generate biological innovations because such technological interventions are based on introducing new genetic information in a particular environment, whether that is a cell or a DNA sequence. Such innovations could be termed technobiological or even technonatural. They are, in the final analysis, canonical examples of the technosciences (see chapter 1).

tion of the said system and its setting. This latter form of natural innovation is particularly important philosophically, given that it does not derive from an action (or the notion of "introducing" something) but instead from a complex set of interactions in a specific environment.

I will acknowledge this suggestion on the part of Wiktionary and thus speak about the *appearance or emergence of something new to a particular environment*, regardless of whether that environment is an ecosystem, a technological environment, or a social environment. Combining this expression with the previous one, I will therefore offer a new provisional definition:

> Definition 3: *Innovations are interactive processes that generate something new in particular (concrete, determined) environments.*

This definition includes several of the components I have discussed already, with two exceptions: that which refers to the creation of value and that linked to the diffusion of innovation. This latest provisional definition can therefore be amended in the following way:

> Definition 4: *Innovations are interactive processes that create value in generating and diffusing something new in particular environments.*

It might seem that this fourth definition does not apply to natural innovations because values make up a sphere that is more properly associated with humans. However, in the next section I will show that there are natural values, such as life, survival, biodiversity, and growth (with the latter understood as an increase in capacities). The emergence of the atmosphere, for example, was something valuable for the living beings that benefited from it and damaging to those that it harmed. It was an innovation that generated both good and bad things. As I noted previously, this is one of the characteristics of the philosophy of innovation I am presenting here.

As regards the Wikipedia article on innovation, it begins like this:

> Innovation is the creation of better or more effective products, processes, services, technologies, or ideas that are readily available to markets, governments, and society. Innovation differs from invention in that innovation refers to the use of a better and, as a result, novel idea or method, whereas invention refers more directly to the creation of the idea or method itself. Innovation differs from improvement in that innovation refers to the notion of doing something different.[28]

This definition is not valid for natural innovations so I cannot use it as such, although it does contain some interesting aspects that are worth discussing. First, it serves for conceptual innovations because it acknowledges that innovation can consist of new ideas. Second, it distinguishes innovation quite well from invention, requiring an invention to be used before it can be termed an innovation (this is one requisite, although one would have to add others). It also distinguishes between innovation and improvement, which is important. Some definitions of innovation require the new thing to be an improvement on its older counterpart, but there is no reason why this should be the case, at least for all the stakeholders involved. A weapon of mass destruction, for example, has a greater capacity to kill than previous weapons, but its use can be prohibited precisely because of that: because of its excessively deadly capacity. As regards the biosphere, the appearance of a new virus through mutation and after having become resistant to antibiotics is a biological innovation that can be lethal for the body or the organ in which the virus lives. They are the setting or the environment, and the mutation of the virus implies an innovation, to be sure, but in no way would one think that this innovation is good for the environment in which it has been produced. My philosophy of innovation is pluralist, which implies opposing evaluations for the same innovation, and also consequentialist: it is not enough to just analyze the end result of an innovation process; one must also examine later consequences for the environment in which the innovation took place. We thus come to a key point. Throughout this work I have repeatedly come across three conceptions of innovation: one based on the agents that promote innovation; another based on the results that are ultimately obtained; and a third,

28. "Innovation," in Wikipedia, accessed August 17, 2012.

which I support, that besides including the other two understands innovation as a process. This process of innovation has several points: that it stems from the initiative of some agents; is developed through time; takes place in a particular environment; is diffused throughout other environments and, where appropriate, scales; is intervened in by diverse agents; obtains different results; has results that are evaluated in different ways by the agents involved (stakeholders); and has results that are evaluated differently according to the value systems that each of the agents involved has and the benefits or hindrances that the innovation process implies for them.

These eight points reflect my way of thinking quite well, but they are still not enough for defining innovation. They are valid for many other human activities such as scientific research or a judicial trial. But what is missing is the most characteristic feature of innovation: novelty. The same is true of the Wikipedia definition I just examined: the term "new" is not mentioned and the phrase "better or more effective" is used instead. The latter term in particular, "effective," has clear technological and business connotations. It appears to suggest that all innovation must imply some form of progress, something that is simply not true: there are innovations that bring about profound regressions, such as practicing a new form of torture. Many researchers on the topic maintain an accumulative and progressive conception of innovation, perhaps because the linear model was based on these kinds of optimistic hypotheses that continue to be invoked. Indeed, a distinction is typically made between accumulative and disruptive innovations, so that it becomes clear that, throughout the history of innovation, there are moments of rupture without any possibility of continuation. Nor are the later consequences of an innovation process typically taken into account. I am referring here, for example, to the risks for or the impact on the environment in which the innovation process takes place. Given that innovations are usually beneficial for some and harmful for others, it is difficult to sustain that, generally speaking, an innovation means an improvement in regard to what existed previously.

We thus come across a problem that I am still to explore in

some depth: for whom is an innovation beneficial (or more effective) and for whom is it not? Or, more still: for whom is something new and for whom not? I will return to this issue in the following sections. Whatever the case, I would conclude that the Wikipedia "definition" has aspects that remain open to criticism although it also contributes some interesting ideas. It is not easy to assimilate natural innovations with human innovations (whether technological, business, social, political, artistic, cultural, and so on). Specifically, there is much confusion over the double meaning of the term, according to how it is applied to a process as whole or only its end results. As my aim is to construct a conceptual framework that is sufficiently general to include different forms of innovation (including natural innovations), I will attempt to overcome that ambiguity by introducing a neologism, in other words, a conceptual innovation that will allow us to distinguish between innovative processes (and actions) and their results. I will term the latter *novations*, reserving the term "innovation" for the actions that generate those results and, more generally, for innovation processes (understood as a set of innovative actions). This is a question of a purely heuristic distinction that can be accepted. To my mind, it allows one to eliminate the confusion and ambivalence that so often prevails. I propose this distinction, in part, so that research into innovation might be more general, productive, and precise. It is just a proposal and is in no way a case of any essentialist affirmation. With time, I will experiment to test whether this conceptual proposal achieves acceptance or not and, therefore, if it is transformed into conceptual innovation of more or less widespread use. With this addition, then, we come to nine distinctive points about the concept of "innovation" that I will now synthesize in the form of definitions, that is, my two heuristic definitions:

> Definition 5: *Innovations are interactive processes that generate something new and valuable in particular environments.*

> Definition 6: *Novations are the results of these innovations.*

I will briefly comment on why such a distinction makes sense. There are innovations propelled by someone (a subject, agent, group, collective, organization, and so on) and others that are not (those that emerge within a system). In both cases there are derived results (new products, new processes, new species, new forms of organization, and so forth) that are novations. These occur in nature (a new star, a new galaxy, a new particle, a new species, and so on) and not just in societies.

Because they are processes, innovations take place in time and space. For that reason, their expansion (spatial diffusion and temporal duration) must be analyzed, as well as any eventual retraction or extinction. Those processes are usually multi-agent in nature. In other words, they are promoted by and affect or impact diverse agents (stakeholders). Figuring out the diverse number of agents involved in an innovative process (or value chain or network) is one of the first tasks for studies of innovation. Acceptance of novation as something new and valuable on the part of a diverse number of agents is a necessary condition for something to be considered a novation, of whatever type this may be. As regards innovation processes, many of them do not arrive at their end; they do not produce tangible results. For that reason, they can be considered as failures. For research on the subject, the study of failed innovations is as important as that of their successful counterparts because it offers much extra knowledge on the issue, especially regarding the obstacles to innovation in a given environment. Put another way, in studies of innovation it is worth applying the symmetry principle of the strong programme associated with the sociology of knowledge. In philosophical terms, a falsifiability methodology must be applied. The interactive processes I mention here can be of a very different nature (for example, mental, linguistic, social, economic, artistic, and so on). The notion of novation attempts to include invention, and specifically conceptual invention, as one of its forms. It allows one to distinguish between innovation and renovation. Renovation is a process of improving a novation but one must not confuse it with an innovation. That is how I intend to resolve the deficiency I noted in the Wikipedia definition.

Let us now check how general definitions 5 and 6 – for the moment, the latest results of my own philosophical inquiry – are. Let us see if they are capable of integrating the diverse definitions of innovation that I have considered in this work.

1. They are of course valid for conceptual innovations (new ideas) because both ideas and definitions stem from interactive processes that take place among neurons, with the environment being a neuronal network and the end result the idea, represented by its word and corresponding mental representation.
2. They are also valid in order to analyze the search for new definitions (or for new data). In this case, different (and in general, mathematical) terms interact with one another, combining and recombining until they offer a new definition, a new piece of data, a new formula, or a new mathematical representation. Typical examples of such processes include systems of equations, resolution, and the search for solutions. In the final analysis, the environment in which new definitions emerge is also the human brain, but diffusion is produced via representations of thought (conversations, presentations, scholarly publications, formulas, data tables, and so on). One could say much the same for artistic and musical works, those in which creative and innovative thought is also expressed via representations. Definitions 5 and 6 can be applied, without any problem, to processes of artistic or musical creation, always on the assumption that a novation (something new) is the consequence, as the end result, of such processes. Note that this new thing does not necessarily need to be a particular work; it can also be a process or an interaction. Postmodern theories about artistic processes fall within my definitions without any particular problem, as do traditional theories based on the notion of a work of art and even a masterpiece. Assigning relevance to innovation processes is also very fruitful in the cultural and creative domain.
3. I would, therefore, say that definitions 5 and 6 comprise various forms of conceptual innovation without any problem. This includes artistic innovations, which reflect specific environments (a canvas, a studio, a concert hall, a city, a performance, and so on), even though their source, par

excellence, is the human brain. Their diffusion, however, is produced via diverse means (languages, sign systems, databases, artistic bases, and so on).

4. This pair of definitions is also valid when it comes to analyzing natural innovations, so long as we understand that living things (such as bacteria) that generate something new (like free oxygen) interact with one another in such cases and that this innovation is then diffused throughout other natural environments (right through to generating the atmosphere itself, as in the aforementioned example). We might also consider inert agents as well as living agents. A volcano is typically a hugely innovative agent, capable of generating whole new islands (such as the Macaronesia series of archipelagoes in the Atlantic Ocean, including the Azores, the Canary Islands, and Cape Verde), but also of destruction and devastation. The examples of floods, tsunamis, and earthquakes are also valid here. In certain cases value is created for some (and destruction for others, including the loss of life). But interactive processes in particular environments are produced and generate (as well as destroy) value, in this case natural value, such as life. For some predators, great natural catastrophes are unique opportunities.
5. Both definitions include those of the *Oslo Manual* as a particular case without any problem. In this case, the environment is the marketplace and processes of interaction are produced within a company, but they're also produced outside companies, with competitors, regulators, consumers or users, suppliers, and, more generally, what in the economy has come to be termed (following Porter) a value chain. The manual identifies four forms of innovation (product, process, organizational, and marketing), but as we already saw in the previous section, all four also make sense in natural environments. Likewise, in chapter 3, we saw how they could also be applied to social innovations. Market innovations are obviously promoted by certain agents and are the result of intentional actions, even

though in these cases, too, there is some margin of chance, that is, uncertainty. They also generate value, and in this case it is countable economic value for investors or on the stock market. And they destroy value, as Schumpeter makes a point of noting. As regards their diffusion, it takes place in other markets, that is, economic (or business) environments. There is no problem in considering the diffusion phase as a stage in the business innovation process, in which interactions with other agents multiply. Markets are, in this sense, multi-agent environments.

6. The same is true of the definitions of social innovation I have discussed to this point. In this case, the interactions are not economic but social; in other words, they are based on the system of social relations that make up a particular social environment. In this case, too, the interactions are multiple and the results must be considered as valuable for social novations to exist.
7. Definitions 5 and 6 also comprise innovation processes that may stem from the cultural and creative sector, without any problem. The interactions, agents, environment, and values change, but this definition embraces them without any special theoretical difficulty. I previously alluded to artistic creations, and my considerations there are valid for other creative practices.
8. Finally, user innovation is also included without any difficulty, especially when users organize themselves and generate user communities or establish communicative links with the remaining agents in the value chain.

In sum, the diverse forms of innovation I have discussed throughout this work, being many and varied, easily fit into definitions 5 and 6, transforming these into particular cases for the very definitions. The most complicated task consists of integrating natural, technological, and user innovations into the same conceptual framework. I believe I have done this insofar as it is possible, even though the definitions I propose here can obviously be improved on.

Ontology of Innovation

I am considering innovation to be an object of study in which, given its complexity, I have detected several distinctive points: it is developed through time; offers new things, shapes processes; generates value; emerges out of actions and interactions; frequently fails; and so on. I will now examine the ontology of innovation more closely before moving on to its axiology, which constitutes the core of the philosophy of innovation I am outlining. Specifically, I will analyze three ontological characteristics of innovation that appear in the heuristic definition I have just proposed: processes, emergence (or generation), and the new.

The expression "innovation processes" is common to many studies of innovation. It is very fitting, and this is why I have also used it on many occasions. One thing is sources or origins, another is the processes that lead to innovation, and another still is the end results. The process/result dichotomy is particularly important, as already pointed out beforehand and as I will demonstrate in what follows.

There are many definitions of innovation, both among researchers of the subject and among its practitioners. However, a definition of the term "process" is rarely put forward. This is hardly surprising because it is an innate notion to living beings (life processes) and, moreover, because it is an abstract and complex notion that incorporates several variants, including mental, physiological, physical, chemical, geological, biological, cosmological, social, economic, legal, technological, musical, and artistic processes. Many things that happen in the world can be understood as processes, beginning with life and followed by thought. Aging is also a process whose end result is death, but ageing and death are two very different things even though one leads to another.

Defining the notion of process is an inescapable conceptual challenge. If innovation is demonstrated in the form of processes, it is not enough to just define innovation – one also has to delimit the compound notion of innovation processes. Curiously, the classic European philosophers paid no attention to this category, except very rarely. The philosophy of processes, which is termed processism or process philosophy, was cultivated mainly by twentieth-century American philosophers such as Charles S. Peirce, William James, Nicholas Rescher, and so on. That said, this particular current of thought was actually initiated in Europe by Henri Bergson and Alfred North Whitehead in the early twentieth century. Reinterpreting Bergson, Whitehead formulated the apothegm "nature is process,"[29] making reference to Heraclitus (everything flows) and Leibniz (monadology) as historical antecedents of process philosophy. Whitehead sets up processism against atomism. Likewise, Leibniz criticizes an atomistic conception of nature. These thinkers believe that natural processes are more important than atoms, which is why processes are for them more important in ontological terms. Rescher is a contemporary proponent of this idea and, in the last two decades, has demonstrated that process philosophy calls into question much of the Western philosophical tradition. According to Rescher, processism contributes a completely different approach to that of classical European metaphysics, whether of the Platonic, Aristotelian, or Kantian variety. Plato observed that authentic reality resided in the world of ideas. Aristotle, meanwhile, centered his thought on substances, and therefore on things, some of which possess a substratum, substance. For Aristotle, this was the object of research for "first philosophy" (metaphysics). And Kantian metaphysics focus on the notion of the thing in itself. None of these thinkers reflected on the ontological category of "process," even though we come across this over and over again, both in the external world and within our very selves: physical and chemical processes, biological processes, physiological and perceptive processes, mental processes, life processes, and so on. For my part, I would add one thing to Rescher's argument: namely, Aristotelian categories include different forms of expressing "being" but not of "becoming." Processes have to do with becoming and change, although they involve things, subjects, and beings that appear to be lasting (substances). Elaborating a logic of becoming is a difficult task, as the Hegel's followers found out. There are deep-seated reasons for this problem, at least if we understand logic from the basis of the subject/predicate relationship: ontologically speaking, a process is not a subject nor is it a

29. See Whitehead, *The Concept of Nature*, chap. 3.

predicate, for all that it might be considered a subject or object of study. Our languages are shaped profoundly by the categories of subject and predicate, and in order to reflect on processes, there is no alternative for us but to use phrases with those grammatical categories. I even recognize that I am speaking of processes *as if they were objects of study*, but I do so because I have positioned myself from a metaphilosophical perspective that objectifies and reifies processes in order to be able to discuss and analyze them. Right now, as I write these lines, I am placing the term "process" in the position of subject, or predicate, again and again. However, as a philosophical category, the notion of process differs greatly from those of object, subject, and predicate. This is the main reason why it is far from easy to elaborate a logic of becoming or of processes.

Having made these qualifications about my way of understanding the philosophy of processes, I will assume a metaphilosophical approach and apply a definition to the category of process, which will in turn have to be added to the previous heuristic definitions of innovation. I will take Rescher as my guide in order to set out on this inquiry. Among contemporary philosophers still alive today, it is Rescher who has perhaps most reflected on processes from an ontological perspective. In one article, Rescher argues, "For process theorists, becoming is no less important than being – rather the reverse."[30] Recalling Heraclitus, Rescher notes that "for him, reality is not a constellation of things at all, but one of processes . . . the river is not an object, but a continuing flow . . . [and] everything is a matter of process, of activity, of change (*panta rhei*)."[31] Processism starts from a metaphysical thesis by which reality, not things, shapes processes. Things are, in fact, the results of processes and, where appropriate, the same as objects and subjects. As regards Leibniz, Rescher contends that he was the main proponent of this line of thought in the modern era, above all, because according to Leibniz, "all the 'things' that figure in our experience (animals alone grudgingly excepted) are mere phenomena and not really 'substances' at all; the world in fact consists of clusters of processes he calls monads (units) which are the 'centers of force' or bundles

30. Rescher, "The Promise of Process Philosophy," 144.
31. Ibid.

of activity; for Leibniz, processes rather than things furnish the basic materials of ontology."[32] In short, the ontology of innovation must be based on an ontology of processes, rather than things, objects, or subjects, even though things, objects, subjects, and other more or less stable entities interact in such processes. This philosophical option I choose here has one important consequence: because processes are ontologically prior to objects, things, and subjects, *the most significant innovations will be those of process rather than those of product*. Rescher does not actually say this because he does not address innovation studies, at least as far as I know. However, I believe such a conclusion concurs with his way of thinking and, in general, with the process philosophy.

In order to develop this notion of process philosophy, Rescher proposes some different metaphysical categories to those of Aristotle and Kant: specifically, those of time, change, contingency, emergence, novelty, and creativity. Clearly, these are all significant when it comes to innovation studies. I will not discuss his ontological proposals, which contribute important conceptual innovations, in great detail, but I will note his central thesis: "For the process philosopher, process has priority over product — both ontologically and epistemically."[33] Although this thesis has multiple philosophical implications, I will limit myself to extracting that which most directly affects innovation studies: namely, *process innovations are prior to (and more important than) product innovations*. Through this work I have shown that product innovations (those carried out by a company) are at the theoretical core of the *Oslo Manual* (in its 1992 edition). Indeed, they still are among economists and political scientists of innovation, with few exceptions, despite that in later editions of the *Oslo Manual*, the OECD and Eurostat acknowledge other forms of innovation (including that of process innovation).

If we accept Rescher's theses and the consequences that, in my opinion, derive from them, the dominant paradigm in innovation studies would suffer a considerable overhaul. The main priority would be to detect the emergence of innovative processes,

32. Ibid.
33. Ibid.

for example new activities, new values, or, to put it in the terms of Amartya Sen, *new capacities*. Later one would have to analyze, define, and measure those innovative processes, more than their end results. And yet measuring processes is no easy task, even though one can establish a temporal pattern for them, distinguishing different phases and stages of the process under consideration. With processes being so difficult to measure, it is much easier to measure the results of an innovation process, for example in terms of the resulting products or product novations. Process philosophy applied to innovation studies must focus on innovation processes more than their results (novative products). This is one of the reasons why, in the previous section, I distinguished between innovation and novation. Novations are results, while innovations are processes. The philosophy of innovation I propose concerns itself with both things, as well as inquiring into sources of innovation, but it emphasizes the intermediary stage — that between origins and results: the processes of innovating, put more accurately.

The principles on which the philosophy of innovation is based, following Rescher, are (in his own words) the following:

It seems sensible to understand 'process philosophy' as a doctrine committed or at any rate inclined to certain basic propositions:

1. That time and change are among the principal categories of metaphysical understanding,
2. that process is a principal category of ontological description,
3. that process are more fundamental, or at any rate not less fundamental than things for the purposes of ontological theory,
4. that several if not all of the major elements of the ontological repertoire (God, nature-as-a-whole, persons, material substances) are best understood in process terms,
5. that contingency, emergence, novelty, and creativity are among the fundamental categories of metaphysical understanding.

A process philosopher, then, is someone for whom temporality, activity, and change — of alteration, striving, passage, and novelty-emergence — are the cardinal factors for our understanding of the real.[34]

In my opinion, though, it would be better to speak in terms of "uncertainty" rather than "contingency." The results of innovation processes are typically uncertain in the same way as their development (which is almost never predetermined) is. Uncertainty and innovation are associated for several reasons, but here I will limit myself to just noting one: different agents take part throughout the course of an innovation process, and it is quite typical that some of them favor the innovation in question and others are opposed to it, due to the fact that its potential implementation will be beneficial for some and damaging for others. This happens in nature, in societies, and in markets, so that we can add uncertainty to the points associated with innovation processes. As regards the other three categories indicated (emergence, novelty, and creativity), I spoke about creativity in the previous chapter, and I will refer to the remaining two in this very section a little later. In another article, Rescher offers a definition of "process" that I will also adopt, although with a few qualifications:

Nothing particularly new or unusual will be said here on the issue of understanding what a process is. A process will be construed in pretty much the usual way — as a sequentially structured sequence of successive stages or phases. Three factors accordingly come to the fore:

1. That a process is a complex — a unity of distinct stages or phases. A process is always a matter of know this, know that.
2. That this complex has a certain temporal coherence and integrity, and that processes accordingly have an ineliminably temporal dimension.
3. That a process has a structure, a formal generic patterning in virtue of which every concrete process is equipped with a shape or format in that its temporal phases exhibit a fixed pattern.

34. Ibid., 144–45.

Although processes themselves are always temporal, they can in general be given an atemporal representation.[35]

As with uncertainty, so temporality and complexity are two other characteristics of processes, including innovation processes. Rescher adds more distinctive features that are of interest: A process must have a structure and a certain integrity to the extent that we can distinguish it from other processes by means of its specific pattern. A firm may set several innovation processes in motion at the same time, but it must develop and evaluate them differently, singling out each process and paying specific attention to it. Moreover, in contending that every process must have a structure, an important line of inquiry is opened up for the philosophy of innovation: the structural changes an innovation causes, if indeed it causes any, must be explored. Innovation processes are produced in specific systems (physical, chemical, biological, business, social, political, and so on) that all have a particular structure. This must be analyzed with the aim of detecting the potential structural changes an innovation generates. In mentioning structure, the possibility is opened up of considering changes in structure like organizational changes, which allows us to revisit one of the four major forms of innovation in the *Oslo Manual*. Above all, one must resort to systems theory, analyzing the components, relationships, and functions in a specific system in order to study the processes of change in that same system. The systemic model of innovation, with the diversity of agents (stakeholders) that intervene in an innovation process, thus acquires an ontological dimension.

The American philosophers who have been at the forefront of process philosophy (Peirce, James, Rescher, and so forth) came to a definition of process that, in my opinion, will be useful according to how it is applied to innovation studies. According to them, a process is a *sequentially structured sequence of successive phases or stages*. Rescher maintains that a sequence with no structure is not a process,[36] so that both the terms "sequence" and "structure" are crucial to the definition of "process." As regards the phases or stages, ultimately they are subprocesses. Therefore, I stand by the ar-

35. Rescher, "On Situating Process Philosophy," 37.
36. Ibid.

gument of processes as systemic entities that are developed in time and whose structure includes subprocesses termed phases, stages, sequences, courses of action, or something else. Rescher constantly underscores the strictly temporal condition of processes, even though they may be represented in atemporal terms, for example by means of mathematical and computational models.

This observation is important because it coincides with what I already established for innovation processes and studies of these processes. The former are practices, actions that are assembled with one another; the latter involve theoretical and methodological work based on the application of conceptual and measurement tools (definitions, indicator systems, statistical data, and so on). Take the example of the computer simulation of a physical, biological, or mental process, as is quite typical nowadays. A computer program is not a process but its execution implies sequences of operations with numbers, symbols, data, images, and so on. Such programs serve to represent diverse processes (for example, physical, chemical, cosmological, biological, and even mental or corporeal). For that reason, we will have to take care not to confuse processes and their representations. In the latter, a process is taken as an object of study. This is what researchers and studies of innovation do. In contrast, practitioners activate innovation processes, which for them are not objects of study but sequences of actions, the exact outcome of which one does not know.

The main methodological and epistemological problem of innovation studies consists of the fact that the *Oslo Manual* paradigm is incapable of representing, quantifying, and measuring processes. It just analyzes their inputs and adds up certain outputs. There are no computational tools to represent these processes, with their phases and stages, while they take place. Studies of innovation should face up to the challenge of developing tools of this kind in the years to come. Focusing solely on inputs (origins, sources, initial resources) and outputs (end results, derived consequences) means disregarding the most interesting part of innovation studies: its condition as a process. Exploring the underlying structure of any innovation process is an unavoidable methodological requirement, as I pointed out in previous chapters: stakeholders need to be determined; their relationships, links, and interactions studied; the evaluations that each of them (and not just the initiator) makes about an innovation analyzed; obstacles identified (and not just the innovative agents); and so on.

This is one of the reasons why a naturalized philosophy of innovation can offer a lot to studies of innovation. Starting from the hypothesis that there are natural innovations – such as physical, chemical, geological, biological, and cosmological innovations – and bearing in mind that some of those processes that include a physical or natural transformation have been represented in computational form (technoscientific simulation, which is today quite typical in many laboratories), it seem plausible to start with such representations and, thereafter, note down and characterize the moments during which some innovation is produced (or better still, emerges). Through this methodology one can make progress in generating computational models for technological, economic, and social innovation, and perhaps also models to analyze creativity and conceptual innovation.[37] Innovation studies do have some qualitative and quantitative tools at their disposal (questionnaires, surveys, data on innovative firms and organizations, statistical models, and so on), but they have still not generated computational or mathematical representations of these processes, just their results and inputs. It is a question of a challenge that opens up the possibility of innovation sciences, should they come into existence. As of now, however, even the dominant paradigm (the *Oslo Manual*) is not very scientifically developed, so that one cannot state that there are any *innovation sciences* as of yet. This is the program to be promoted in innovation studies and a naturalized philosophy can offer much in this regard.

The strength of computational models resides in their potential to represent, via software, certain physical, biological, or other processes. Despite being complex, computational processes are operable with computers and allow one to represent more complex systems and processes, analyzing their sequences of stages. The way I see it, one should begin with those models and introduce

37. In fact, the cognitive sciences already use tools of this kind in order to examine creativity.

into them procedures designed to detect the emergence of innovations, some of which will be successful and others not. In short, it is a question of marking out, in computational form, the emergence of a novation throughout processes that have been computer simulated. The specific goal would be to detect, by means of a computer, when a novation is produced in a given environment, on the basis of a prior operative and computational (that is, heuristic) definition of what a novation might be in such an environment. One should remember that the founding metaphor of evolutionary economics compares firms to biological entities. From a naturalized perspective, already existing models for biological evolution processes would have to be implemented in order to allow one to interpret and represent innovation processes – in principle, in a natural environment and later in other settings, including competitive markets or other social areas. From the plural perspective I propose, these markets are some environments for innovation processes, but they are not the only environment and sometimes not even the main one. In sum, process philosophy questions the traditional conceptual framework of Western ontology. The same thing happens with the philosophy of innovation processes, which differs profoundly from studies based on the analysis of results or novations. It is perhaps now more understandable why I differentiate between a philosophy of innovation and another of novation. The former is a philosophy of processes and the latter a philosophy of subjects, objects, and things, based on inputs and outputs. There is, then, a radical opposition between the two, although this may not seem so at first glance. I propose *entering into the black box of innovation*, that is, analyzing the structure and dynamics of the systems in which innovations emerge, whatever kind of systems and innovations they are.

Rescher contends that every process is a structured, and even sequentially structured, sequence. That is equivalent to saying that the structure in question is iterative and that some subprocesses are repeated over and over again, as happens in computer languages, which function with multiple loops. Bear in mind that the book you are reading, and which is now coming to an end, is the result of successive phases of compiling, translation, and correction that have been carried out by several people (the author, translator, copy editor, designer, indexer, and so on) with the aid of computational tools. Conceptual innovations, if they take place at all during these phases in putting the book together, will have been generated by some cognitive process on the part of the author or another agent in the chain. Yet this is not the important thing, nor is it that of the book as a printed object; instead what is important are the mental processes that this book may induce among you, the readers. That is where user innovation, which is the most important thing, resides. For that reason, the notion of process is defined as a sequentially structured sequence: because it generates new repetitions of basic sequences, and that is what its structure consists of, its specific nature as a process. And the writing/reading duality is a good example of a sequentially structured sequence.

Nevertheless, the example I am giving and others that could be offered call into question the purely sequential characterization of processes. Indeed, there are many processes in which interactions and retroactions are produced among their diverse phases and stages: this has been the case in the compiling of this book, part of which is made up of previous publications by the author and of previously unpublished texts that have now been reworked and published. I believe the same is true of the interpretations or commentaries that could be made about the work. This repetition of steps occurs in most innovation processes, including user innovation, even when they are extremely different. If we only think about a user innovation that generates new products, the prior and continued use of previous products is a condition sine qua non for the emergence of the innovation, as Eric von Hippel has demonstrated on repeated occasions. Each user undertakes his or her own sequence of actions when it comes to using, for example, a technological product; and all users carry out similar sequences, which is why sequentially structured sequences are produced. According to how these actions—which are sometimes not only iterated but also monotonous—are carried out, the human mind can notice possible improvements in that sequence of actions, and therefore also in the corresponding product. User innovation emerges in a context like this, even though it may then have to be diffused, which makes up a later phase. However, the same user does not usually just carry out sequences of actions, as if he or she were a machine. That user also reflects about, imagines, and recalls similar actions; exchanges experiences and knowledge with other people; and so on. All these loops break the sequential structure and give rise to a more complex structure: a *multisequential* structure, to put it one way. Therefore I will adopt, with some caution, the characterization of processes as sequences. As a minimum, these would be multisequences, which include loops and retroactions among some of the sequences involved. There are events in the past that, even though they have already taken place, continue to exert an influence on the course of current events, and therefore on the future as well. This is a very typical case in innovation systems, in which the prior existence of one or another structural conditions favors (or hinders) the development of these innovation processes. The notion of sequence seems too linear to accept it as a theoretical base for studying innovation processes. In some cases, interactive processes of the kind that emerge in innovations may be entirely sequential and be carried out through consecutive phases, but on other occasions interactions are inter-temporal because subprocesses and actions carried out in the past have an influence on them. Imagining the future also plays an important role in an innovation process. In situations like this it is not enough to just speak about the temporal nature of processes, although this is true; one must take a step further and allude to the inter-temporal nature of the multi-agent environments in which human (and natural!) innovation processes take place.

This is my main qualification of the definition formulated by Rescher. That does not stop me from adopting this same definition in order to analyze innovation processes, at least initially, and as a heuristic tool widely accepted by several philosophers. As regards the definition, it is sufficiently precise, general, and fruitful, so that it adapts well to the epistemic values that I have selected to guide my inquiry. It is also valid as a heuristic definition in order to move forward in elaborating a philosophy of innovation because it analyzes one of the aforementioned points in the definition of innovation I adopt here: namely, that of *an interactive process that generates something new and valuable in a particular environment*. One of the most complex components of this definition is the notion of *process*. Another point that is no less complex is that of value, especially in the case of natural innovations. I will address this in the next section. Before moving on to that, I will say a couple of words about the other two points I make use of in the definition: "generation" and "new." With regard to the former, I prefer to speak about *emergence* (of something new, following the interaction process) and sometimes *appearance*, which is more illustrative in some cases. On other occasions innovations are produced, that is, they emerge according to a prior design and as the consequence of a set of intentional actions. This is the case of many business, technological, and political innovations as well as, more generally, those innovations produced by human beings. However, the vast majority of natural innovations do not stem from intentional actions. I therefore prefer to speak about appearance or emergence, with the latter being a more general term. One should also speak about survival or about unexpectedly occurring properties in a specific system.

A market analyst is capable of detecting the emergence of a new product in a particular market and analyzing the consequences that its appearance generates, even though that new product is the result of an innovation process that a specific firm has developed. The innovator carries out intentional actions in order to do so, but what happens in the market is the appearance of a new product: in other words, the emergence of something new. This terminology is also valid for innovation processes that take place in nature, which is why I will speak about emergences and survivals (of new products, new processes, new forms of organization, and new forms of marketing and communication) that occur on the basis of interactive processes in which several agents (rather than just one) intervene. One should recall, too, that after the emergence of an innovation it must be consolidated and diffused, whether in a commercial, social, or natural environment. The diffusion of innovations is one of the most significant phases in any innovation process and many agents intervene in that stage. In order to conclude this section, I will offer some brief considerations about the concepts of *emergence* and *diffusion* because both form part of the definition I have adopted, even though they do not appear explicitly in that definition.

Few philosophers have explored the concept of emergence. One of them is Mario Bunge, a staunch defender of an atomistic ontology based on the notion of things and objects, although he also takes the existence of processes into account. According to him, "A new thing possessing an emergent property is sometimes called an *emergent*. And the process whereby a thing loses one or more properties may be called *submergence*."[38] Previously, Bunge defines the notion of an emergent property in the following terms: "A property of a complex object is said to be *emergent* if neither of the constituents or precursors of the object possesses it."[39]

These definitions of emergence and emergent properties are not particularly useful for my own inquiry because they identify emergence along the lines of something new and nothing else. In principle, all emergences imply novelty but the reverse is not the case. Emergent properties appear in a system following certain prior interactions among the components of that system and bring with them a change to that system as a whole or at least to parts of it. A new property can be added to previous ones; an emergent property, however, affects the whole set of properties in a system, modifying several of them, although a new property also appears. Therefore I will not adopt Bunge's definition of emergence.[40]

Steven Johnson, meanwhile, offers a very different definition: "The movement from low-level rules to higher-level sophistication we call emergence."[41] And to illustrate this definition, he introduces the concept of complex behavior by describing "a system with multiple agents dynamically interacting in multiple ways, following local rules and oblivious to any higher-level instructions."[42]

38. Bunge, *Emergence and Convergence*, 17.
39. Ibid.
40. Nevertheless, what Bunge then goes on to say is significant: "Things are not the only possible emergent: processes too may emerge and submerge." Ibid. We thus return to the idea I already discussed as regards Rescher: the most significant innovations are those that generate new processes, not just new things (products, objects).
41. Johnson, *Emergence*, 18.
42. Ibid., 19.

Although Johnson's definition is not very precise in conceptual terms, the underlying idea seems to me fitting when it comes to characterizing the notion of emergence, given that it links emergence to complex systems. Innovations, whether natural, technological, economic, social, or artistic, emerge thanks to multi-agent systems that interact with each other in multiple ways; out of this combination of interrelations, on certain occasions, something new appears, whether a rule, a product, a process, or a property. The "new" is related to "emergence," but the two notions should not be confused. For the latter to take place, those multiple prior tests are necessary (for example, Darwin's variations or symbiogenetic combinations), which must then be checked against the environment in such a way that some of them disappear and others keep on struggling to survive and consolidate themselves in the corresponding environment. In other words, emergences require an internal interaction within a complex system, and also an interaction with the external environment. Because multiple agents intervene, the success of some combinations and failure of others cannot be predicted; this is only decided empirically, and with possible loops and retroactions through time. In order to illustrate what I am talking about, it is worth citing the example that Johnson himself uses later on:

> The neural networks of the brain are densely interconnected: each individual neuron contains links—in the form of axons and synapses—to as many as a thousand other neurons. When a given neuron fires, it relays that charge to all those other cells, which, if certain conditions are met, then in turn relay the charge to their connections, and so on. If each neuron extended a link to one or two fellow neurons, the chance of a reverberating loop would be greatly reduced. But because neurons reach out in so many directions simultaneously, it's far more likely that a given neuron firing will wind its way back to the original source.[43]

These interactions are typical in user innovation, as are retroactions, so that the subprocesses return to their origin and are repeated in different ways, with new variables, until some of its variants are successful and generate something new. These inter-

43. Ibid., 134.

actions are also produced in innovations stemming from R&D, although in this case fewer agents intervene in the innovation process. Above all, that complex system of interactions always takes place in nature, both in terms of Darwinian and symbiogenetic interactions, given that in one or another interaction they are produced with multiple agents and with an environment in general, and on repeated occasions. A natural novation will never happen the first time around.

I trust these brief considerations might serve to clarify, at least in part, the concept of emergence that I use here; still, I acknowledge that this concept is extremely difficult to define strictly. It should be made clear that I am only referring here to the emergence of novations after an innovation process and not to a more general theory of emergence in any system whatsoever.

If we now move on to consider the notion of novelty, with the aim of differentiating it from the notions of innovation and of emergence, we will see that this is also not a straightforward task because they are all interrelated notions. Jon-Arild Johannessen, Bjørn Olsen, and G.T. Lumpkin list three issues related to novelty or newness: "In order to isolate a useful definition and measure of innovation, we need to address three newness-related questions: what is new, how new, and new to whom?"[44] They refer exclusively to business innovation, but these three questions also make sense when it comes to a general theory of innovation, although qualifying the third point somewhat. In the case of nature, there can be innovations without any subject one might perceive as such, so that the third part of the question, for my purposes, should be framed thus: where and when do innovations appear? The environment forms part of the definition of innovation; it is not something external to it, because the very notion of newness is relative to what existed beforehand. In a particular system, of whatever type, something is new in relation to what there used to be in that system in its previous state, before the introduction, appearance, or emergence of the "new" was produced. In other words, novelties are always contextual, which is why the place and the time in which they are produced must enter into any characterization

44. Johannessen, Olsen, and Lumpkin, "Innovation as Newness," 21.

of the new; ultimately, nothing is new without some reference to what came beforehand.

One more reason to deny any essentialist definitions of innovation and affirm its strictly pluralist condition is that innovations are always relative to the space and the time in which they are produced. In other words, both innovations and novations are "situated." It is only possible to work out what is new from the basis of an empirical study of the environment in which such an emergence occurs. It is also worth underscoring the fact that innovation is a question of degrees, as pointed out by Johannessen, Olsen, and Lumpkin. There are major innovations (of rupture or disruptive) and small innovations (accumulative, of improvement). Put another way, innovations usually appear in microcosms and are then diffused in space and time. That is why they make up processes that are sometimes developed and other times die out. One must not confuse the origin of an innovation (its invention) with the process that allows for its implementation and diffusion in space and time. Like the new day, innovations gradually expand to the different areas of where they appeared. The diffusion phase of innovations is just as or more important than the appearance phase (that is, its thinking up, design, invention, and so on) because throughout it interactions among multiple agents, who either accept or reject the proposed innovation, are produced. Most innovation processes fail during this diffusion phase and dissolve without generating any tangible result. Not all innovations give rise to novations. It is the same in natural environments, so that one can argue that the existence of a diffusion phase forms part of the structure of an innovation process and, therefore, the definition of innovation.

Before focusing on the diffusion phase it is worth commenting on the response that Johannessen, Olsen, and Lumpkin give to the first of their questions: what is new? They identify up to six different classes of the new: new products, new services, new methods of production, opening new markets, new sources of supply, and new ways of organizing.[45] From the perspective of an ontology of innovation, this classification can be improved on because it is designed exclusively for one type of innovation: that of business in-

45. Ibid.

novation. From a natural perspective, one should not speak about products or services but about objects (things), subjects, and relations. As regards new methods of production or supply, it is preferable to speak about new processes, in more general terms, and in particular about *new processes of interaction* that may be linked to production and supply, as well as about other kinds of interaction, including those agents that (sometimes actively) oppose and hinder innovation. In these cases too, new actions and interactions can appear. This meaning is more important than the first one – the innovation of objects and things (products) – as I argued previously. In the following section I will link it to new capacities of action, following Amartya Sen, and even new values. From my axiological perspective, the emergence of new values is particularly significant for innovation studies. The fourth type proposed by Johannessen, Olsen, and Lumpkin – opening new markets – is clearly inspired by Schumpeter. It touches on the core of the diffusion phase, but from an ontology of innovation, it is preferable to speak about new environments or new areas for innovation processes because markets make up one environment of innovation but not the only one. As regards the sixth type – new ways of organizing – I accept it as it is, given that it is very important in natural innovation processes. Throughout the history of evolution, many disruptive innovations have emerged out of organizational modifications, both internal (within a living organism) and external (in an ecosystem).

I could mention another type not listed by Johannessen, Olsen, and Lumpkin but one that was introduced by the third edition of the *Oslo Manual*: marketing and organizational innovations. In the chapter on social innovation in the present work, we saw how a more general term for these is communication innovations, which is what I use here. However, communication is just a type of relation among people (or animals), in the same way as commercial relations, so that it is enough to speak in terms of new relations in order to incorporate this final type of novelty. It is an open option.

In sum, I will identify several types of novelty and therefore innovation: new processes, new relations, new objects (things, subjects), new environments, and new forms of organization. Given the definition of innovation I am using, it is inferred that these new forms of organization refer above all to the organization of interactions that take place in an innovation process. Organizational innovations could be included within relational innovations, but it is best to differentiate between them because an organism and an organization imply a specific relational system that is much more complex and systemic than a simple collection of relations.

Having offered these considerations on the notion of novelty or newness, I must now move on to analyzing the diffusion phase of innovations, in which the success or failure of all innovative proposals is figured out, both in nature and markets as well as in societies. In order to do so, I will resort to a classic 1962 work by Everett M. Rogers on the diffusion of innovations.[46] Rogers was a sociologist who brought together over five hundred studies of innovations in the rural sphere, allowing him to establish a theory of diffusion that continues to be a point of reference to this day.

He characterizes diffusion as a process made up of four components: "Diffusion is the process by which an innovation is communicated through certain channels among the members of a social system."[47] A definition like this is valid for both technological and social innovations. However, in the case of natural innovation one would have to replace the term "social system" with that of "ecosystem," and instead of "communication" one would have to speak in terms of "transmission" or another word like that. In short, one would have to say something along the following lines: *diffusion is the process by which an innovation is transmitted in an ecosystem through certain means*. For technological social innovations, meanwhile, that means is usually communication (marketing), but in natural innovations the means of transmission is physical: namely, by means of the air, water, earth, movement, energy, or any other element that allows for interaction among natural agents. As regards time, it is a component that is included in the notion of process, so that diffusion would be a phase within innovation process. Rogers believes that, above all, new ideas are diffused, and he reasons out innovation by means of analogy with scientific practice. For my part, I have already observed that conceptual and epistemic

46. Rogers, *Diffusion of Innovations*.
47. Ibid., 5.

innovations are just one form of innovation, even though they examples par excellence. Rogers also warns that "the newness means that some degree of uncertainty is involved in diffusion."[48] Therefore we can confirm the presence of some degree of uncertainty in any diffusion process, and therefore in any innovation process. He also relates the theory of innovation to that of social change, given that "when new ideas are invented, diffused, and are adopted or rejected, leading to certain consequences, social change occurs."[49] Thus, he is clear on the point that, in the diffusion phase, the most important thing is the consequences that are derived from the innovation, to the extent that this is accepted or rejected by society. As regards the notion itself of innovation, he defines it thus: "An innovation is an idea, practice or object that is perceived as new by an individual or other unit of adoption."[50] The most interesting element in this definition is the requisite of perception: namely, something is new if someone perceives it as such. This warning is central to the case of technological and social innovations. In the case of natural innovations, it would be better to say *if someone experiences it as such*: for example, if a new predator appears in an ecosystem, if an organism develops a new process that allows it to carry out an organic function it already performed in another way (accumulative innovation), or if a new function were to appear in this organism that it did not previously carry out. Rogers does not refer to these types of innovations or to their diffusion because he is only concerned with business innovations, yet for a natural philosophy of innovation they are very important. The emergence of a new (natural or social) function or (technological) functionality within an object, thing, subject, process, or any other type of entity is an innovation par excellence, one that generates new capacities and in general new functions. I will discuss this in the next section.

As a final comment on Rogers's work, he is very clear on the point that the evaluations made in regard to an innovation can be very different in the diffusion phase: "The same innovation may be desirable for one adopter in one situation, but undesirable for

48. Ibid., 6.
49. Ibid., 7.
50. Ibid., 11.

another potential adopter in a different situation."[51] I have emphasized this point on more than one occasion throughout this work, but it is still worth making it again. There are innovations that are favorable for some but damaging to others. The evaluations of an innovation that the different agents concerned (stakeholders) make throughout the diffusion process may be different and even contradictory. For this reason, the evaluations that each agent makes are decisive in order to give value (positive or negative) to an innovation. Each evaluates innovations according to their own value systems. And in the final analysis, these evaluations are decisive in order to know which innovation is favorable for an agent, group, or organization, and which is not. The same is true in the case of natural innovations. The relationship between innovation processes and the value systems that intervene in them is complex, but also decisive for the success or failure of an innovative proposal.

My ontology of innovation is based on systems theory, as I have suggested more than once here. For that reason, one must bear in mind what the basic ontological categories are of this theory, because they will be what allow us to distinguish different "natural classes" of innovation. My interpretation of systems theory, made from an exclusively ontological perspective, involves at least ten types of entities: objects, subjects, actions, relations, functions, transformations, structures, processes, systems, and environments. As well as new objects (things, products, ideas, elements), I will also speak about new subjects (human or not, individual or not, active or passive), new relations (whether social, economic, based on ecosystems, and so on), new actions (activities, interactions, utilizations, appropriations, and so on), new structures (organizations, infrastructures, and so on), new processes (whether communicational, physiological, economic, social, and so forth), new transformations (unexpected changes, mutations, emergences, extinctions, and so on), new systems, and new environments. The emergence of a radical or disruptive innovation usually generates new systems, such as new technological systems, but sometimes also gives rise to new ecosystems and environments.

51. Ibid., 12.

Note how these types of innovation, in principle those ten I just mentioned, include (to put it one way) a certain ontological scaling. The appearance of a new object or thing, like that of a new subject (whether an individual, company, or organization), is undoubtedly an innovation, but these are not disruptive innovations, with some exceptions. The more an innovation transversely affects the ten aforementioned ontological categories, the more profound and radical it is. The Internet is a good example of this kind of radical innovation because it gave rise to not just new objects, subjects, relations, and so on, but also and above all it generated a new technological system (the ICT system of information technologies). And, in the final analysis, the Internet created a new social space as well: cyberspace, the digital world, the electronic space, or the third environment, as I prefer to call it.[52]

In sum, and by way of a conclusion to this section, in my ontology of innovation I identify at least ten "natural classes," given that I acknowledge ten categories in the ontology of systems. It may be the case, of course, that the same process generates several types of innovation: for example, new objects, new relations, and new structures (organizations). The great innovations usually affect all or almost all the ontological categories I identify.

Throughout this work I have insinuated on more than one occasion that one of the great innovations consists of the emergence of *new values*. A typical example is the appearance, consolidation, and diffusion of environmental values in the twentieth century, or from another point of view, the passing of the Universal Declaration of Human Rights by the United Nations in 1948. This declaration implied a disruptive legal innovation in the same way as the environmental movement. In the ontological categorization I have just presented, values are considered as functions. Measurements are prime examples of (in general, quantitative) evaluation, in the same way as indicator systems. In order to conclude my philosophical reflection, I will now focus on the relations between innovation and values — the central theme of this work. I will thus move, in an uninterrupted way, from the ontology to the axiology of innovation.

52. See Echeverría, *Los Señores del Aire*.

The Axiology of Innovation

There is much to be done in the philosophy of innovation. Here I have limited myself to outlining some principal lines of inquiry in order to pose the issue and develop it to some extent. Insofar as one accepts that there is no innovation without a plural number of evaluations of the proposals undertaken – whether these are epistemic, technological, business-related, political, social, artistic, legal, or of any other kind – the different forms of human innovation are based on values. First one must evaluate them as something new and then as something beneficial or prejudicial for the agent concerned.[53] Therefore, it is true that any innovation process gives rise to a diverse number of evaluations that can be convergent but also divergent. According to the different types of value at stake, the procedures to evaluate which innovations are successful and which are not will be different. Innovations are always evaluated but there is no universal regulation for doing so owing to the fact that numerous value systems are involved, some of which contradict one another.

That stated in the previous paragraph is equally valid for those innovations generated by humans, but not for natural innovations. Before considering the possibility of an axiology of innovation, one must address the difficult question of whether values exist or not in nature, and if so, what they are.

For the empiricist tradition, the mere formulation of a question like that would be scandalous. Since David Hume, there has been a philosophical tradition that believes in the radical division between facts and values. It is enough to mention the abundant literature of the twentieth century on the naturalistic fallacy (on the part of G.E. Moore and others), according to which one cannot infer value judgments from fact-based principles. Prior to Hume, Western philosophical thought established that moral, ethical, and axiological questions were specific to human beings. For that reason, in stating that there are natural values, one comes up against

53. In chapter 1, I identified twelve types of values: basic (natural), epistemic, technological, economic, social, political, military, legal, environmental, esthetic, moral, and religious, although one could also speak in terms of cultural values. Therefore from an axiological perspective, there would be just as many forms of innovation.

two great traditions in thought, both of which are deep rooted in modernity. It is true that, as regards the second of these, environmentalism and the pro-animal rights movement have modified the state of the issue. There is no lack of ethologists who attribute certain moral capacities to some superior animals such as primates. This is the case of Frans de Waal, for example, who concludes his book *Good Natured* (1996) in the following terms: "I hesitate to call members of any species other than our own 'moral beings,' yet I also believe that many of the sentiments and cognitive abilities underlying human morality antedate the appearance of our species on this planet."[54] For my part, I agree with these reflections, but I would add that the issue of natural good and bad has little to do with ethical and moral questions because it is linked to another kind of "good" that is much more basic, like life, survival, food, water, air, or shelter, to mention just some of the many goods that exist in nature and whose loss or absence constitutes a "bad" thing, both for a species as a whole and for an individual. Human beings themselves, in that they are animals, radically depend on these goods, which we term necessities. If a person lacks air to breathe, they will be asphyxiated and die. The absence of air is bad and its presence is good, always so long as there is the correct proportion of oxygen, that which is suitable for the human body. Air is not a moral good for human beings but rather a natural good, as any astronaut knows, for whom air is an artificial (that is, technological) good. We could say much the same for many other naturally good and bad things that form part of the so-called kingdom of necessity, which some thinkers have contrasted to the moral kingdom. For my part, axiology must concern itself with all kinds of good and bad things. Axiology is in effect much more wide-ranging than a moral approach.

Therefore, when speaking about natural values I am not trying to inquire into any "natural morality" in the biosphere, but rather I seek to analyze the different good and bad things there are there, which is very different. From this, I will be able to attempt an explanation of whether innovations are naturally good or bad, and for whom. As regards animal rights activists, they are right when

54. Waal, *Good Natured*, 210.

they say that some animals possess the capacity to suffer, as well as that of sacrificing themselves for their group companions, which implies altruistic behavior. Animals also experience non-moral good and bad things in their own bodies, and feel them as such, selecting and seeking out the former and defending themselves against and fleeing from the latter. Still without being rational or moral subjects, their behaviors are guided by good and bad things, and therefore by values.

In effect, I believe that where there is good and bad there are values. In fact, I will define such good and bad in the following way:

> Definition 7: *Good is a value that concerns (or is attributed to) an object. One should also say that the object satisfies this value.*

> Definition 8: *Bad is a countervalue (a negative value) that concerns (or is attributed to) an object. One should also say that the object satisfies this countervalue, or that it dissatisfies the opposing value.*

Staring with these definitions, it is clear that innovation is also a value, of course in an economic sense but also in social, artistic, epistemic, and technological ways (among others). If some people consider it good (or bad), then this is because they evaluate innovations positively (or negatively), in other words, because they apply axiological functions to innovation processes and also to the end novations. However, I would never say that innovations, considered in general terms, are morally good or bad. Some may be (for example, a new method of torture), but speaking in general terms, innovation is not a moral value. There may be exceptions: for example, when someone invents and practices a new moral behavior, such as Gandhi and his hunger strikes. But there are few cases like this, and they are not the rule but the exception. The axiology of innovation I will present is not part of ethics, although they do coincide in some ways.

Similarly, there is natural good and bad, and therefore there are also natural values, some of which are very important like biodiversity, sustainability, or balance in an ecosystem. However, those values are not moral but ecological or, if one prefers, environmental. In contrast to many theoreticians of environmentalism who consider their contributions to be an extension of ethics, I contend that environmental values and natural values form two special classes of values that should not be confused or mixed up with ethical values, for all that there may be interrelations among certain natural, environmental, and ethical values. Much the same could be said for the epistemic values (precision, rigor, generality, empirical suitability, productiveness, and so on) I mentioned earlier when discussing Kuhn. They are not moral but epistemic or cognitive values because they are linked to our cognitive capacities and not to our moral capacities. This does not mean that moral values do not exist in scientific activity, for example the value of not altering the data obtained in an experiment, but epistemic and moral values are two different types of value. And I would say the same thing for other types of values.

I will now examine in more detail the plural nature of those values and their relations with good and bad, with the goal of better understanding in which sense I consider innovation to be a natural value and not just a technological, economic, social, epistemic, or artistic value. I will sum up my approach in the form of six points:[55]

1. In specific linguistic contexts, the terms "novelty" (or "newness") and "innovation" function like values. One only need witness media competition for an exclusive story to justify my argument regarding all things new. As regards innovation, the official documents of different governments (for example, those of the European Commission) state over and over again that innovation is the driving force of the economy because it generates growth, well-being, and employment. Therefore that word does not just possess positive value connotations but it would also appear to have been transformed into the solution for many current problems. Innovation and other economic values (competitiveness, productivity, efficiency, and so on) guide public economic policies and business strategies,

55. For a more extensive discussion of these ideas, see Echeverría, *Ciencia y Valores* and *Ciencia del bien y el mal.*

including innovation policies. In maintaining and prioritizing these axiological theses, certain types of values are favored because they are considered more important and are therefore prioritized to the detriment of all others. Proponents of this idea would say that only the kind of innovation that creates economic value is of any value. One must recognize that this is typical, but also that there are other forms of innovation, based on other values, including natural and biological innovations. If a particular animal species generates a mutation that allows it to run more or swim better or fly faster, and if that mutation is consolidated genetically and it adapts to a particular environment (diffusion), the possibilities for that species surviving and reproducing increase. This is good for the species, but not in a moral sense. Instead, it is good from the perspective of natural values, in which survival and reproduction are good.

2. The same thing happens in social sectors, which differ from their economic counterparts: scientists' actions are guided by epistemic values, those of engineers by technological values, those of artists by esthetic values, those of politicians by political values, those of jurists by legal values, and so on. Environmentalists, for their part, orientate their actions on the basis of ecological or environmental values and soldiers affirm military values (victory, dignity, discipline, and so forth), as one would expect. The axiology of innovation must take account of and analyze the different value systems that refer to the search for newness and the urge to innovate, in whatever social sector they may be. Axiologically, innovation processes are plural.
3. This allows one to identify different types of innovation on the basis of axiological criteria. It also opens up a way of measuring the degree to which different values are satisfied by means of evaluation matrixes, including innovation insofar as it functions as a value. Evaluation matrixes are the outcome of the evaluation processes of human actions and, in this case, innovative actions. When the level

of satisfaction of particular values can be measured, those matrixes are numerical and one can function with them. The axiology I propose includes axiological functions, some of which are metrical and not just quantifiable. The search for indicator systems for social innovation heads in that direction.

4. In order to specify exactly the concept of value, the most direct route consists of extending the ontology of Gottlob Frege, who in the early twentieth century originated mathematical logic. Just as Frege conceives of concepts as unsaturated functions that are applied to objects,[56] thereby breaking with the Aristotelian tradition that considers them predicates of a subject, the same idea can be applied to values. According to this hypothesis, values are the result of applying *axiological functions* to any kind of entity, object, event, or process, with those functions possessing different properties according to how they are applied to these different entities. There are cases in which functions can be added up, and even multiplied, thereby generating evaluations that can be mathematized. In other cases this is not possible, and thus one must proceed by means of evaluative sentences. Moreover, axiological functions can always be combined and compounded in such a way that one is applied first and another thereafter. This is decisive when it comes to representing different evaluations throughout an innovation process. We saw previously that, in order to figure out whether an activity is innovative or not, one must successively apply several criteria. Well, this is one of the formal characteristics of functions: the possibility of being compounded sequentially. In principle, it is possible to represent different evaluations of an innovation via functions. The category of process is important for innovation studies in the systems theory I have adopted, but so is that of function because it encompasses different axiological functions, including the metrical functions that are indispensable for operational definitions of innovation.

56. See Frege, "Function and Concept."

5. One should not confuse values with value judgments or with expressing preferences. A value judgment linguistically expresses an evaluative thought, which in turn is the result of applying an axiological function or a set of axiological functions; for example, when, after having weighed up many things, one states that *this innovation process has been beneficial (positive) for this firm*. Someone utters that principle, but the evaluation is prior to the statement in that it stems from an individual or collective evaluation process from which a person who is responsible for stating the value judgment (a manager, boss, evaluation agency, department of in industry, expert user, and so on) establishes a dictum. However, before the final evaluation is formulated, different agents have applied a set of evaluation criteria to the novations, which have led to obtaining that principle or value judgment as an end result. The *Frascati Manual* evaluates R&D in a similar way, and the same thing happens in the evaluation of scientific research projects and papers presented for publication in specialized journals. Successive evaluations, in line with how a process unfolds, are very typical. Innovation processes are accompanied by evaluation processes for the simple reason that the term "innovation" operates nowadays like a significant axiological function.
6. The different agents that evaluate in an innovation system possess an evaluative (or judgment) capacity. In other words, they are enabled to apply axiological functions to objects (agents, systems, instruments, and so on) until a specific end result is obtained that may be qualitative (good or bad, on the cardinal scale), comparative (better or worse, on the ordinal scale), or quantitative ("x" amount of points, on the metric scale). Values are the result of the act of evaluating, with the latter understood as the application of an axiological function v on a variable x, that is, on any x objects so long as they can be evaluated via the v function. From this perspective, stating that a specific activity is an innovation process implies applying the ax-

iological function "i" (innovation) to a set or structured sequence of actions, whether human or not. Axiological functions also exist in natural environments, although in this case they are exemplified by bodies and their natural conducts (the struggle for survival, search for food and water, defense and protection of offspring, collaboration with animals from the same species, and so on). From the axiological perspective I propose, the term "innovation" is a value, that is, an axiological function that is applied to processes and activities. For a bovine herd, finding new fields in which to graze or new sources of water to drink from are innovations guided by the value of survival; these can be compared to discovery of new markets on the part of the Schumpeterian entrepreneur. A business or social activity could be innovative or not according to evaluation criteria applied. The *Oslo Manual*, for example, constructs an international standard in order to classify an activity as innovative or not in a market economy environment. This standard defines and makes operative (in other words, measures) what the axiological function "i" is, as well as a way of applying it. For all that some of aspects of this might be criticized, the "Oslo model" continues to be canonical in scholarly studies of innovation. Here I interpret it as an operative axiological function or set of functions that make it possible to discern and measure innovations in the private business sector. In other environments, the functions to identify, evaluate, and measure innovation will be different because they start from different value systems.

One should not think of values as ideal entities or atemporal essences. *Values are temporal and contextual, given that they emerge as the result of applying axiological functions to different types of variables (objects, relations, processes, and so on)*. Put another way, values only exist when they are embodied in things, people, objects, or processes, that is, when the spheres of value and being mutually overlap one another. The same thing happens in innovation, which is not an idea but a process or set of processes. In general, values are demonstrated as objects, things, processes, or, from a formal point of view, variables that do not exist per se in that they are formalizations. Furthermore, the same value can change over time and according to the contexts in which it is applied (evolutionary axiology).

If this happens with the value of "innovation," something similar will happen with other economic values that are complementary or connected to innovation: profitability, productivity, competitiveness, efficiency, and so on. These axiological functions have different connotations in each market, whether global, national, international, regional, or local. Nevertheless, although the application of axiological functions may be different according context and era, their numerical results are comparable with one another and also inter-operable. This is the crux of the question: to define innovation operationally. This is why scientific studies of innovation are possible and not just literature on innovation. Axiology becomes science to the extent that values are expressed, classified, compared, and measured through scientific methods and concepts, and in my case, via *evaluation matrixes* that are the result of applying measurement protocols to processes, activities, agents, and so on. Although as individuals we all conceive values in our own way, and if in the act of evaluating we use common protocols, the results we each obtain will be comparable and, generally, open to being mathematized. Using the same measurement instruments is one of the bases of the scientific method, and it allows the resulting data to be comparable. In the case of innovation studies, when international agreement was reached over a standard procedure and indicator system in order to measure the phenomenon, a set of axiological functions (evaluation criteria) began to be compared that, in the end, allowed for the figuring out of what innovation is (operationally speaking) and what it is not. The *Oslo Manual*, for all its faults and deficiencies, demonstrates that the task is possible. It is a similar objective when it comes to promoting *scientific studies of social innovation*. In order to work out an alternative paradigm, one must manage to get the resulting data to be able to represent and compare different innovation systems. In the best-case scenario, one should be able to compare human societies and social species, although this is a difficult goal to achieve. For that reason, one should limit oneself to a philosophy of innovation without having to speak about a general science of innovation.

My observations here are not just valid for economic values but for other types of value as well. Business innovation is guided by the intention of increasing the productivity, competitiveness, and efficiency of a company; in the final analysis, its goal is to improve economic benefits, however these may be measured. One could say much the same for innovations in the public sector insofar as they tend to improve or extend public services, reaching wider levels of the population. In such cases, one must pay attention to public opinion surveys, election results, user evaluations, and so on. These quantitative data offer a representation of the axiological function that allows one to measure the level of success of a political proposal: whether it is innovative or not is something that the people themselves must work out.

The same rule applies to social innovation, at least in principle. It is necessary to identify which socially good things innovative agents seek to promote as well as the socially bad things they try to deny, diminish, or resolve. Those new ideas that, once developed and diffused, increase socially good things and diminish their bad counterparts in a particular social environment will be candidates to be considered as the seeds of social innovation in that environment. The values to be promoted are specific to each of many societies, as are the socially bad things to be combated.

In order to develop innovation, however, it is not enough to boost positive values by means of new ideas and initiatives. This is a necessary condition but not sufficient in itself. The main thing consists of specifying exactly what the axiological functions are that derive from those values, as well as the formulas and protocols in which they take shape. This is the function of evaluation matrixes, which weigh up and measure the degree to which each of the values, and each subsystem of values, is met. By way of a summary, the conceptual framework in which my axiological approach to innovation is located is the following:

1. Human ideas are guided and oriented by values, without being determined by them.

2. The innovative activities promoted by a company, institution, or organization need not have precise objectives, and may even be chance attempts, as occurs in the case of biological innovations in natural evolution. However, these actions are guided by values in the same way as innovation processes in nature. Axiological rationality consists of increasing any available good and lessening any bad. It is an intrinsically pluralist rationality because the good and bad things in question depend on a particular environment and the different agents that interact with one another there.
3. There are natural values (survival, nutrition, mobility, health, and so on) that, to paraphrase Polanyi, could be termed tacit values or also corporeal values: they orient our actions even though we are not conscious of them because they are *constitutive values of our very selves*, that is, of our bodies or, where appropriate, they are *constitutive values of the ecosystems and environments we are examining*. Something similar happens with languages, with knowledge in R&D systems, and with productivity, profitability, and competitiveness in the case of innovation. Contrary to what positivist thinkers believe, R&D systems, like innovation systems, are full of value, and moreover of a very diverse nature: epistemic, technological, economic, political, military legal, social, cultural, environmental, and, last but not least, esthetic, religious, and moral.
4. The axiological conception of rationality states that it is possible to evaluate ends, and not just means. Indeed, nobody defines a goal for their actions without having previously evaluated various objectives, accepting some as possible, others as desirable, some as utopian, and others as questionable or too predictable. *Teleological instrumental rationality* is based on *axiological rationality*, precisely because any goal, objective, or end that a human being proposes can be evaluated ex ante and ex post in positive or negative form.[57] The innovating imperative never deter-

57. For more on this notion of bounded axiological rationality, see Álvarez and Echeverría, "Bounded Rationality in Social Sciences."

mines which actions must be undertaken. Innovation processes are almost always developed in conditions of uncertainty. However, the "innovation" value shapes company strategies as well as those of certain public institutions and social organizations.

5. There is no one dominant universal value, not even life, and much less wealth or economic development. The notion of growth as a maximum value has been criticized, and people have now begun to speak in terms of *sustainable development*. Analogically, it would be a grave error to consider innovation as an end in itself, or as a value that predominates over all others. Social innovations must be accepted by society, and thus it is necessary for them to be evaluated positively. This may not be the case, of course, and they may be rejected. I would say the same thing about other forms of innovation.

If we understand values as axiological functions, it so happens that the values are limited among themselves because it is impossible to maximize the level of satisfaction of each of the axiological functions. Axiological pluralism has the virtue of limiting the maximum levels of satisfaction of values, precisely because one must satisfy several criteria of evaluation rather than just one. Innovation systems are axiologically pluralist owing to the fact that different agents, whose actions are guided by different and even opposing values, coexist within them. It is impossible to be a maximum innovator because innovation processes set different agents, each of whom guides their actions on the basis of diverse values, against one another. The aim of maximizing a value usually fails because there are other values to satisfy and not all values are subordinate to the others. Trying to be the most innovative can in the long-term lead to delirium.

The axiological proposals I have just made are improved if they are combined with the theories of Amartya Sen on the spaces of capabilities and functionings. Sen proposes these as a technical alternative to the rational decision-making theories maintained by utilitarians, according to which every human being is characterized as an economic agent via his or her space of utilities and behaves rationally when he or she attempts to maximize his or her utility function. The spaces of capabilities and functionings are, though, based on a plurality of values, not just on utility:

> The concept of "functionings," which has distinctly Aristotelian roots, reflects the various things a person may value doing or being. The valued functionings may vary from elementary ones, such as being adequately nourished and being free from avoidable disease, to very complex activities or personal states, such as being able to take part in the life of the community and having self-respect.

A person's "capability" refers to the alternative combinations of functionings that are feasible for her to achieve.[58]

And later on, Sen adds:

> The amount or the extent of each functioning enjoyed by a person may be represented by a real number, and when this is done, a person's actual achievement can be seen as a *functioning vector*. The "capability set" would consist of the alternative functioning vectors that she can choose from. . . . The evaluative focus of this "capability approach" can be either on the *realized* functionings (what a person is actually able to do) or on the *capability set* of alternatives she has (her real opportunities). The two give different types of information — the former about the things a person does and the latter about the things a person is substantively free to do.[59]

Therefore, Sen affirms the existence of a plurality of values that acknowledge different levels of satisfaction. Throughout one's life, a person tries to satisfy, to some extent, the values he or she believes to be significant. However, there are occasions on which a human being's space of capability (what he or she can choose to do) is very restricted because it is forced by necessity and that person barely has any freedom to act. In such cases, Sen speaks about poverty, and even misery, and contends that it is possible to measure this by means of functioning vectors and capability vectors. Consequently, Sen offers a new measurement system for poverty and wealth, understood from a pluralist axiological perspective. In this regard, he is very explicit: "The capability perspective is inescapably plu-

58. Sen, *Development as Freedom*, 75.
59. Ibid.

ralist."[60]

The axiology I propose incorporates these ideas of Sen's, but it extends them in two directions: on the one hand, I apply the notions of spaces of capabilities and functionings to collective agents and not just to individuals, as Sen does. On the other, I extend his proposals to the animal world, contending that every living thing possesses its own capabilities of action and interaction with the environment, and that such capabilities define their own natural values. I will not expand in any great detail on either of these additions to Sen's theories because it would lead me off my main theme here. I will instead just limit myself to suggesting an interpretation of natural evolution in terms of capabilities and functionings, with both understood as forms of interacting with an environment.

In a situation like this, then, natural innovations produce something very similar to that produced by technological or social innovations: namely, they transform the space of capabilities of each agent, whether animal or human, individual or collective, or also private or public in the case of organizations. The introduction of new technology into a company, an office, or a house modifies the capabilities of action of those who work or live in these environments. Therefore the space of capabilities is transformed, and sometimes radically. Another thing is whether those new capabilities are translated or not into functionings. In order to do that, one must learn to use new technologies, acquire skills and abilities, evaluate the results of new actions, share knowledge and experiences with other users, and so on. Whatever the case, technological innovations are linked to the development of new capabilities of action and interaction. The example of the Internet clearly demonstrates this in that it facilitates relations and interactions among humans over great distances and online.

This is evident in the case of technological innovations but it is also valid for social as well as natural innovations. When a living species develops the capacity to fly, for example, its space of capabilities experiments a radical change: this is a disruptive innovation that affects both its spaces of capabilities and functionings. Because these depend on the natural values germane to that spe-

60. Ibid., 76.

cies, the ultimate effect is the emergence of new values or, at the very least, new forms of applying previously existing values. One could say much the same for individuals: learning to walk takes a lot of effort and causes a lot of bruises or scratches, yet once one has learned how to do so, the two spaces – of capabilities and functionings – are transformed considerably. The important thing is the capacity to walk more than the existence of paths. As the Spanish poet, Antonio Machado, wrote, "*Caminante, no hay camino, se hace camino al andar*" (Walker, there is no path, the path is made through walking).

The same could be said about collective and institutional agents, which also possess their own (institutional, shared, company, and so on) values, opportunities, limitations on acting and choosing (capabilities), and effective functionings. Sen underscores the fact that the space of capabilities and functionings enjoys many advantages in this regard when compared to the space of utilities, based on the individual function of utility.[61]

From my perspective, values are axiological functions that individual or collective agents apply in order to value objects, people, events, instruments, and so on. One can evaluate one's own lifestyle, for example, together with one's own life projects, that is, one's current life functionings and space of life capabilities. Values shape the decisions that each and every one of us make. For that reason, functionings can be considered like effectively applied axiological functions whose results are valued by subjects themselves. Put another way, values are reflected empirically in functionings or operations. As regards the space of capabilities, it is linked to the different options that one may choose at a specific given moment or throughout one's life, which also implies prior valuations. Sen's focus encompasses very effectively my own conception of values as axiological functions that are satisfied gradually, as well as the starting point of axiological pluralism.

This focus is also valid for collective agents and consequently for organizations. An institution, company, political party, NGO, and many communities are characterized by sharing determined valued, that is, by collectively embracing a certain value system,

61. Ibid.

even though that might be with certain qualifications or different levels of acceptance. Each member of a collective will apply those values in a different way when it comes to guiding their actions, but there is also a common axiological core to the actions of those different agents that sustains *collective actions* and the *comparison between the actions of one another*. This is what teamwork consists of, from my axiological perspective.

Thus a collective agent can be characterized by a value system that guides (to a greater or lesser degree) its actions. On the basis of this, it is possible to define, however vaguely, the space of operations of that same collective agent (what an institution or organization does) as well as its space of capabilities (what it can do). If we add the notion of axiological function to Sen's approach, we are thus in a position to address, from a new perspective, one of the major problems of the social sciences: collective action. Intentions do not add up together (Kenneth Arrow), and therefore the intentional theory of action has experienced great difficulty when analyzing collective action. However, if actions are guided by values that are made up among themselves and can be shared by various agents, the notions of collective agent and collective action can be analyzed. In conclusion, Sen's ideas, reinterpreted in this way, allow us to formulate an *axiological conception of action* that, at least in principle, is valid for both individual and collective agents of very diverse kinds.

With all this, we are in a better position to address the problem of innovation as a value. In effect, two typical kinds of innovation are those that generate new relations and those that generate new actions; this is something that happens thanks to the introduction of new technology such as the telephone or the automobile, but also as a consequence of formulating the technical protocols of interconnection and driver's manuals. Whether technological or social, innovations impact on the space of functionings, both in the case of people and of organizations. For that reason, they can in principle be measured by means of the techniques proposed by Sen in order to measure poverty, although of course in my axiology other types of values would have to be applied. The evaluation matrixes to which I have already referred in this work and in previous publications make up the most suitable instrument for the axiology of innovation I propose here.

Those matrixes allow one to jointly evaluate different types of values, given that human actions are not just guided by one single value but instead by a plurality of values, which are integrated to one extent or another but are reconcilable as functions. In the event of there being several agents with different values, as is typical in the diffusion phases of an innovation, then one has to measure the level of satisfaction of each of the agents involved in the innovation process. This is what is done in sociological surveys when those surveyed are asked to score specific options on a scale of one to ten. In the event of innovations being measured, one should measure the impact of the innovation in question on the level of satisfaction of the remaining values, and all this for each of the agents involved (the stakeholders). In sum, the axiology of science can have at its disposal the tools of empirical analysis used to measure not only novations (the end results of an innovation process) but also the effects and consequences of these processes on their direct promoters and other agents concerned with them.

BIBLIOGRAPHY

Adorno, Theodor W. "Culture Industry Reconsidered." Translated by Anson G. Rabinbach. *New German Critique* 6 (Fall 1975): 12–19.

Adorno, Theodor W., and Max Horkheimer. "The Culture Industry: Enlightenment as Mass Deception." In *Dialectic of Enlightenment: Philosophical Fragments*, edited by Gunzelin Schmid Noerr, translated by Edmund Jephcott. Stanford: Stanford University Press, 2002.

Agassi, Joseph. "Why There Is No Theory of Models." In *Theories and Models in Scientific Processes*, edited by William E. Herfel, Wladyslaw Krajewski, Ilkka Niiniluoto, and Ryszard Wójcicki. Amsterdam: Rodopi, 1995.

Agazzi, Evandro. *La techno-science et l'identité de l'homme contemporain: Quelques raisons de la crise d'identité de l'homme contemporain*. Fribourg: Ed. Universitaires Fribourg Suisse, 1997.

———. *Right, Wrong and Science: The Ethical Dimensions of the Techno-Scientific Enterprise*. Edited by Craig Dilworth. Amsterdam: Rodopi, 2004.

Aho, Esko, et al. *Report of the Independent Expert Group on R+D and Innovation*. Luxembourg: European Commission EUR 22005, 2006.

Allee, Verna. "Value Creating Networks: Organizational Issues and Challenges." *The Learning Organization* 6, no. 6

(2009): 427–42.

Álvarez, José Francisco. "Capacidades potenciales y valores en la tecnología: elementos para una axionomía de la tecnología." In *Ciencia, Tecnología, Sociedad y Cultura en el cambio de siglo*, edited by José A. López Cerezo and José M. Sánchez Ron. Madrid: Biblioteca Nueva/OEI, 2001.

Álvarez, José Francisco, and Javier Echeverría. "Bounded Rationality in Social Sciences." In *Epistemology and the Social*, edited by Evandro Agazzi, Javier Echeverría, and Amparo Gómez Rodríguez. Póznan Studies in the Philosophy of Science 96. Amsterdam: Rodopi, 2008.

Andari, Robert, Hasan Bakshi, Will Hutton, Áine O'Keeffe, and Philippe Schneider. *Staying Ahead: The Economic Performance of the UK's Creative Industries*. London: Department for Culture, Media and Sport, 2007. www.theworkfoundation.com/assets/docs/publications/176_stayingahead.pdf.

Andersen, Kristina Vaarst, Høgni Kalsø Hansen, Arne Isaksen, and Mika Raunio. "Nordic City Regions in the Creative Class Debate: Putting the Creative Class Thesis to a Test." *Industry & Innovation* 17, no. 2 (2010): 215–40.

Arboníes, Ángel L. *¿Innovación o evolución? Metáfora evolutiva de empresa*. Madrid: Díaz de Santos, 2007.

Arundel, Anthony. "Innovation Survey Indicators: What Impact on Innovation Policy." In *Science, Technology and Innovation Indicators in a Changing World – Responding to Policy Needs*. Proceedings of the OECD Blue Sky II Forum, Ottawa, 25–27 September. Paris: OECD, 2007.

Arundel, Anthony, and Hugo Hollanders. *EXIS: An Exploratory Approach to Innovation Scoreboards*. Brussels: European Commission; DG Enterprise, 2005.

———. *2006 European Innovation Scoreboard Methodology Report: Searching the Forest for the Trees – "Missing" Indicators of Innovation*. Brussels: European Commission; DG Enterprise, 2006.

———. *2006 Trend Chart Methodology Report: Searching the Forest for the Trees: "Missing" Indicators of Innovation*. Maastricht:

MERIT – Maastricht Economic Research on Innovation and Technology, 2006.

———. *Innovation Union Scoreboard IUS*. Maastricht: UNU-MERIT, 2010.

———. *A Taxonomy of Innovation: How Do Public Sector Agencies Innovate? Results of the 2010 European Innobarometer Survey of Public Agencies*. Maastricht: UNU-MERIT, 2011.

Axelrod, Robert. "Advancing the Art of Simulation in the Social Sciences." In *Simulating Social Phenomena*, edited by Rosario Conte, Rainer Hegselmann, and Pietro Terna. Berlin: Springer, 1997.

Bakshi, Hasan, and Eric McVittie. "Creative Supply Chain Linkages and Innovation: Do the Creative Industries Stimulate Business Innovation in the Wider Economy?" *Innovation: Management, Policy & Practice* 11, no. 2 (2009): 169–89.

Bakshi, Hasan, and David Throsby. *Innovation in Arts and Cultural Organizations*. NESTA Interim Research Report. London: NESTA, 2009.

Bakshi, Hasan, Eric McVittie, and James Simmie. *Creating Innovation: Do the Creative industries Support Innovation in the Wider Economy?* NESTA Research Report. London: NESTA, 2008.

Bakshi, Hasan, Philippe Schneider, and Christopher Walker. *Arts and Humanities Research and Innovation*. London: Arts and Humanities Research Council; NESTA, 2008.

Blaug, Mark. "Where Are We Now on Cultural Economics." *Journal of Economic Surveys* 15, no. 2 (2001): 123–43.

Bloch, Carter. *Measuring Innovation in the Nordic Countries: Copenhagen Manual*. Nordic Innovation Centre (NICe) Project No. 08207. Copenhagen: Danish Center for Research and Research Policy, 2011.

Bloch, Carter, and Vladimir López-Bassols. *Innovation in Firms: A Microeconomic Perspective*. Paris: OECD, 2009.

Boekholt, Patries. "The Evolution of Innovation Paradigms and their Influence on Research, Technological Development and Innovation Policy Instruments." In *The Theory and Practice of Innovation Policy*, edited by Ruud E. Smits, Stefan Kuhlmann, and Philip Shapira. Cheltenham, UK: Edward Elgar, 2010.

Borins, Sandford. "Encouraging Innovation in the Public Sector." *Journal of Intellectual Capital* 2, no. 3 (2001): 310–19.

Braaksma, R. M., J. P. J. de Jong, and E. Stam. *Creatieve bedrijvigheid in Nederland*. Zoetermeer, Netherlands: EIM, 2005.

Breton, Philippe. *Historia y crítica de la informática*. Translated by Víctor Villoria. Madrid: Cátedra, 1989. Originally published as *Histoire de l'informatique* (Paris: La Découverte, 1987).

Buchwald, Jed Z., ed. *Scientific Practice: Theories and Stories of Doing Physics*. Chicago: University of Chicago Press, 1995.

Bugge, Markus M., Peter S. Mortensen, and Carter Bloch. *Measuring Public Innovation in Nordic Countries: Report on the Nordic Pilot Studies – Analyses of Methodology and Results*. Rapport 40/2011. Oslo: Nordic Institute for Studies in Innovation, 2011.

Bunge, Mario. *Emergence and Convergence: Qualitative Novelty and the Unity of Knowledge*. Toronto: University of Toronto Press, 2003.

Bureau of European Policy Advisers (BEPA), European Commission. *Conclusions of the BEPA Workshop on Social Innovation*. D(2009). Brussels: Bureau of European Policy Advisers, 2009.

———. *Social innovation as Part of the Europe 2020 Strategy*. Executive Summary. Brussels: Bureau of European Policy Advisers, 2009.

———. *Empowering People, Driving Change: Social Innovation in the European Union*. Report prepared by Agnès Hubert, Matteo Bonifacio, and Joep Konings. Brussels: Bureau of European Policy Advisers, 2010. http://ec.europa.eu/bepa/pdf/publications_pdf/social_innovation.pdf.

Bush, Vannevar. *Science, the Endless Frontier*. Washington, DC: United States Government Printing Office, 1945. www.nsf.gov/od/lpa/nsf50/vbush1945.htm.

Castells, Manuel, ed. *La sociedad red: Una visión global*. Madrid: Alianza, 2006.

Castells, Manuel, et al. *La transición a la sociedad red*. Barcelona: Ariel/UOC, 2007.

Caves, Richard E. *Creative Industries: Contracts between Art and Commerce*. Cambridge, MA: Harvard University Press, 2000.

Chesbrough, Henry W. *Open Innovation: The New Imperative for Creating and Profiting from Technology*. Boston, MA: Harvard Business School Press, 2003.

———. *Open Business Models: How to Thrive in the New Innovation Landscape*. Cambridge, MA: Harvard Business School Press, 2006.

Cloutier, Julie. *Qu'est-ce que l'innovation sociale?* Les cahiers du Crises. Collection Études théoriques, ET0314. Québec: Crises, 2003.

Conseil de la Science et de la Technologie (CST). *Innovation sociale et innovation technologique. L'apport de la recherche en sciences sociales et humaines*. Québec: Les Publications du Québec, 2000.

Culture Statistics Program. *Canadian Framework for Culture Statistics*. Research Paper. Catalogue no. 81-595-MIE2004021. Ottowa: Culture, Tourism and the Centre for Education Statistics Division, 2004.

Cunningham, Stuart D. "From Cultural to Creative Industries: Theory, Industry, and Policy Implications." *Media International Australia Incorporating Culture and Policy: Quarterly Journal of Media Research and Resources* (2002): 54–65.

———. *What Price a Creative Economy?* Platform Papers 9. Redfern, N.S.W., Australia: Currency House, 2006.

———. "The Creative Economy: Patterning the Future." *Dialogue: Academy of the Social Sciences in Australia* 26, no. 1 (2007): 15–23.

Cunningham, Stuart D., and Peter L. Higgs. "Measuring Creative Employment: Implications for Innovation Policy." *Innovation: Management, Policy and Practice* 11, no. 2 (2009): 190–200.

De Jong, Jeroen P. J., and Eric von Hippel. *User Innovation in SMEs: Incidence and Transfer to Producers*. Scales Research Reports

H200814. EIM Business and Policy Research. Zoetermeer, Netherlands: Scales, 2008.

Dees, J. Gregory. "The Meaning of 'Social Entrepreneurship.'" Center for the Advancement of Social Entrepreneurship (CASE), Duke University, 2001.

Dennis, Michael. "A Change of State: The Political Cultures of Technical Practice at the MIT Instrumentation Laboratory and the John Hopkins University Applied Physics Laboratory, 1930–1945." PhD diss., John Hopkins University, 1990.

Department for Business, Innovation and Skills, Government of the United Kingdom. *Digital Britain*. Norwich: TSO, 2009.

Department for Culture, Media and Sport (DCMS), Government of the United Kingdom. *Creative Industries Mapping Document*. Creative Task Force. London: DCMS, 1998.

———. *Creative Industries Mapping Document*. 2nd ed. Creative Task Force. London: DCMS, 2001.

———. *Creative Industries Economic Estimates Statistical Bulletin*. London: DCMS, 2007.

———. *Creative Britain: New Talents for the New Economy*. DCMS/ BERR /DIUS. London: DCMS, 2008.

———. *Creative Industries Economic Estimates Statistical Bulletin January 2009*. London: DCMS, 2009.

Department for Culture, Media and Sport and Department of Environment Transport and Regions. *Creative Industries: The Regional Dimension*. The Report of the Regional Issues Working Group. London: DCMS, 2000.

Department for Innovation, Universities and Skills (DIUS), Government of the United Kingdom. *Innovation Nation*. Cm 7345. London: The Stationery Office, 2008. www.bis.gov.uk/assets/biscore/corporate/migratedD/ec_group/18-08-C_b.

Deroïan, Frédéric. "Formation of Social Networks and Diffusion of Innovations." *Research Policy* 31, no. 5 (2002): 835–46.

Díez, José Antonio, and Ulises Moulines. *Fundamentos de filosofía de la ciencia*. Barcelona: Ariel, 1997.

Dosi, Giovanni. "Technological Paradigms and Technological Trajectories." *Research Policy* 11, no. 3 (1982): 147–62.

Drucker, Peter F. *Innovation and Entrepreneurship: Practice and Principles*. New York: Harper & Row, 1985.

———. "The Discipline of Innovation." *Harvard Business Review* 63, no. 3 (May–June 1985): 67–72.

Echeverría, Javier. *Los Señores del Aire. Telépolis y el Tercer Entorno*. Barcelona: Destino, 1999.

———. *Ciencia y Valores*. Barcelona: Destino, 2002.

———. *La revolución tecnocientífica*. Madrid: FCE, 2003.

———. *Ciencia del bien y el mal*. Barcelona: Herder, 2007.

———. "Towards a Philosophy of Scientific Practice: From Scientific Theories to Scientific Agendas." In *Filosofia, Scienza e Bioetica nel dibattito contemporaneo*, edited by Fabio Minazzi. Rome: Instituto Poligrafico e Zecca dello Staa, 2007.

———. "Aportaciones preliminares a los estudios de innovación social." In *Textos y pretextos para repensar lo social*, edited by Ignacio Mendiola. Bilbao: Universidad del País Vasco, 2008.

———. "La innovación desde una perspectiva filosófica." Innovación sin fronteras: el mito de la sociedad del conocimiento. Monograph 20. *Mi+d*, online journal (2008): 32–36.

———. "Innovar en la sociedad del conocimiento." *Telos* 11 (2008): 22–28.

———. "El manual de Oslo y la innovación social." *Arbor* 732 (2008): 609–18.

———. "De las políticas de investigación a las políticas de innovación." *Acta Sociológica* 55, no. 1 (2010): 13–37.

———. "Technomathematical Models in Social Sciences." In *Probabilities, Laws, and Structures*, edited by Dennis Dieks, Wenceslao J. González, Stephan Hartmann, Michael Stöltzner, and Marcel Weber. Dordrecht, Netherlands: Springer, 2012.

Echeverría, Javier, Andoni Alonso and Pedro J. Oiarzabal, *Knowledge Communities*. Reno: Center for Basque Studies, University of Nevada, Reno 2011.

Echeverría, Javier, and Lucía Merino. "Cambio de paradigma en los estudios de innovación: el giro social en las políticas europeas de innovación." *Arbor* 752 (2011): 1031–43.

Elster, Jon. *Explaining Technical Change: A Case Study in the Philosophy of Science*. Cambridge: Cambridge University Press, 1983.

ESSnet. *Project ESSnet Culture*. Final Report draft. Brussels: European Commission/Eurostat, 2011.

Etzkowitz, Henry, and Loet Leydesdorff, eds. *Universities and the Global Knowledge Economy: A Triple Helix of University-Industry-Government Relations*. London: Pinter, 1997.

———. "The Dynamics of Innovation: From National Systems and 'Mode 2' to a Triple Helix of University-Industry-Government Relations." *Research Policy* 29: 2 (2000): 109–23.

European Commission. *Green Paper on Innovation*. COM(95) 688. Brussels: European Commission, 1995.

———. *Innovation in a Knowledge-driven Economy*. COM(2000) 567. Brussels: European Commission, 2000.

———. *Productivity: The Key to Competitiveness of European Economies and Enterprises*. SEC(2002) 262. Brussels: European Commission, 2002.

———. *Innovation Policy in Europe 2002*. Innovation Papers no. 29. Luxembourg: Office for Official Publications of the European Communities, 2003.

———. *Innovation Policy: Updating the Union's Approach in the Context of the Lisbon Strategy*. COM(2003) 112. Brussels: European Commission, 2003.

———. *More Research and Innovation: Investment for Growth and Employment*. COM(2005) 488. Brussels: European Commission, 2005.

———. *On the Social Agenda*. COM(2005) 33. Brussels: European Commission, 2005.

———. *Putting Knowledge into Practice: A Broad Based Innovation Strategy for the EU*. COM(2006) 502. Brussels: European Commission, 2006.

———. *The Renewed Social Agenda: Opportunities, Access and Solidarity in 21st Century Europe*. Communication from the Commission to the European Parliament, the Council, the

European Economic and Social Committee, and the Committee of the Regions. SEC(2008) 2156, SEC(2008) 2157, SEC(2008) 2178, and SEC(2008) 2184. Brussels: European Commission, 2008.

———. *Elements for the Setting-up of Headline Indicators for Innovation in Support of the Europe 2020 Strategy*. Report of the High Level Panel on the Measurement of Innovation. Brussels: DG Research and Innovation, 2010.

———. *Europe 2020: A Strategy for Smart, Sustainable and Inclusive Growth*. Communication from the Commission COM(2010) 2020. Brussels: European Commission, 2010.

———. *Europe 2020 Flagship Initiative: Innovation Union*. COM(2010) 546. Brussels: European Commission, 2010.

———. *European Competitiveness Report 2010*. Commission Staff Working Document SEC(2010) 1276. Brussels: European Commission, 2010.

———. *Green Paper: Unlocking the Potential of Cultural and Creative Industries*. COM(2010) 183. Brussels: European Commission, 2010.

———. *Creative Europe: A New Framework Programme for the Cultural and Creative Sectors (2014–2020)*. COM(2011) 786/2. Brussels: European Commission, 2011.

———. *Innobarometer 2010. Analytical Report: Innovation in Public Administration*. Brussels: European Commission, 2011.

———. *The New Renaissance*. Report of the "Comité des sages" Reflection Group on Bringing Europe's Heritage Online. Brussels: European Commission, 2011.

———. *Policy Handbook*. Working Group of EU Members States Experts. Brussels: European Commission, 2012.

———. *Promoting Cultural and Creative Sectors for Growth and Jobs in the EU*. Communication from the Commission to the European Parliament, the Council, the European Economic and Social Committee, and the Committee of the Regions. COM(2012) 537. Brussels: European Commission, 2012.

———. *Guide for Social Innovation*. Brussels: European Commission, 2013.

European Commission and PRO INNO Europe/Innometrics. *European Innovation Scoreboard 2007: Comparative Analysis of Innovation Performance*. PRO INNO Europe paper No. 6. Luxembourg: Office for Official Publications of the European Communities, 2008.

European Commission Enterprise & Industry Directorate-General (2010a), New Cluster Concepts Activities in Creative Industries, Simposium "Towards a Pan-European initiative in support of innovative creative industries in Europe", Amsterdam, 4-5 February 2010.

Eurostat. *Cultural Statistics*. Luxemburg: Publications of the European Union, 2011.

Flew, Terry, and Stuart D. Cunningham. "Creative Industries after the First Decade of Debate." *The Information Society* 26, no. 2 (2010): 113–23.

Florida, Richard L. *The Rise of the Creative Class, and How It's Transforming Work, Leisure, Community and Everyday Life*. New York: Basic Books, 2004 (paperback edition).

———. *Who's Your City? How the Creative Economy Is Making Where You Live the Most Important Decision of Your Life*. New York: Basic Books, 2008.

Florida, Richard L., and Irene Tingali. *Europe in the Creative Age*. London: Alfred Sloan Foundation and Demos, 2004.

Flowers, Stephen H., Tanja Sinozic, and Parimal Patel. *Prevalence of User Innovation in the EU: Analysis Based on the Innobarometer Surveys of 2007 and 2009*. Innometrics Thematic Paper. Brussels: PRO INNO Europe/Innometrics, 2009.

Flowers, Stephen, Eric von Hippel, Jeroen de Jong, and Tanja Sinozic. *Measuring User Innovation in the UK: The Importance of Product Creation by Users*. Research Report. London: NESTA, 2010.

Franklin, Allan. *The Neglect of Experiment*. Cambridge: Cambridge University Press, 1986.

Frege, Gottlob. "Function and Concept." In *The Frege Reader*, edited by Michael Beaney. Oxford: Blackwell Publishers, 1997.

Freeman, Christopher. *The Economics of Industrial Innovation*. Harmondsworth, UK: Penguin Books, 1974. — — —. "The National System of Innovation in Historical Perspective." *Cambridge Journal of Economics* 19, no. 1 1995: 5–24.

Frey, Bruno S. *Arts & Economics: Analysis & Cultural Policy*. Berlin: Springer, 2000.

———. "Public Support." In *A Handbook of Cultural Economics*, edited by Ruth Towse. Cheltenham, UK: Edward Elgar, 2003.

Frontier Economics. *Creative Industry Performance: A Statistical Analysis for the DCMS*. London: Frontier Economics Ltd, 2007.

Fundación Española para la Ciencia y la Tecnología (FECT). *Libro Blanco sobre Arte, Ciencia y Tecnología*. Madrid: FECYT, 2006.

Funtowicz, Silvio O., and Jerome R. Ravetz. "Science for the Post-Normal Age." *Futures* 25, no. 7 (1993): 739–55.

———. *La çiencia posnormal: ciencia con la gente*. Barcelona: Icaria, 2000.

Galison, Peter. *How Experiments End*. Chicago: University of Chicago Press, 1987.

———. "Introduction: The Many Faces of Big Science." In *Big Science: The Growth of Large-Scale Research*, edited by Peter Galison and Bruce Hevly. Palo Alto, CA: Stanford University Press, 1992.

Garnham, Nicholas. "From Cultural to Creative Industries." *International Journal of Cultural Policy* 11, no. 1 (2005): 15–29.

Gault, Fred. *Social Impacts of the Development of Science, Technology and Innovation Indicators*. Working Paper. United Nations University and UNU-Merit Document 2011-008. Maastricht: MERIT, 2008.

———. *Innovation Strategies for a Global Economy: Development, Implementation, Measurement and Management*. Cheltenham, UK: Edward Elgar; Ottawa: IDRC, 2010.

———. "Developing a Science of Innovation Policy Internationally." In *The Science of Science Policy: A Handbook*, edited by Kaye Husbands Fealing, Julia I. Lane, John H. Marburger III, and Stephanie S. Shipp. Palo Alto, CA: Stanford University Press, 2011.

———. *User Innovation and the Market*. Working Paper, United Nations University and UNU-Merit Document 2011-009. Maastricht: MERIT, 2011.

Gault, Fred, and Eric von Hippel. *The Prevalence of User Innovation and Free Innovation Transfers: Implication for Statistical Indicators and Innovation Policy*. MIT Sloan School of Management Working Paper 4722-09. Cambridge, MA: MIT Sloan School of Management, 2009.

Gibson, Chris, and Lily Kong. "Cultural Economy: A Critical Review." *Progress in Human Geography* 29, no. 5 (2005): 541–61.

Ginsburgh, Victor A. "The Economics of Art and Culture." In *International Encyclopedia of the Social and Behavioural Sciences*, edited by Neil J. Smelser and Paul B. Baltes. Amsterdam: Elsevier, 2001.

Godin, Benoît. *Innovation: The History of a Category*. Project on the Intellectual History of Innovation. Working Paper No. 1. Montreal: INRS, 2008.

———. *Social Innovation: Utopias of Innovation from* c.*1830 to the Present*. Project on the Intellectual History of Innovation. Working Paper No. 11. Montreal: INRS, 2012.

———. "καινοτομία: An Old Word for a New World; or, The De-Contestation of a Political and Contested Concept." In *Challenging the Innovation Paradigm*, edited by Karl-Erik Sveiby, Pemilla Gripenberg, and Beata Segercrantz. London, Routledge, 2012.

Goldenberg, Mark. *Social Innovation in Canada – How the Non-profit Sector Serves Canadians… and How It Can Serve Them Better*. Canadian Policy Research Networks (CPRN) Research Paper W 25 Work Network. Ontario: Canadian Policy Research Networks, 2004.

Goldenberg, Mark, Wathira Kamoji, Larry Orton, and Michael Williamson. *Social Innovation in Canada: An Update*. CPRN Research Report. Ontario: Canadian Policy Research Networks, 2009.

Gouvernement de Québec, Conseil québécois de la recherche sociale (CQRS). *Recherche en sciences humaines et sociales et innovations sociales. Contribution à une politique de l'immatériel.* Par Camil Bouchard, en collaboration avec le Groupe de travail sur l'innovation sociale (GTIS). Québec: Les Publications du Québec, 1999.

Gouvernement de Quebéc, Conseil de la science et de la technologie (CST). *Innovation sociale et innovation technologique: L'apport de la recherche en sciences sociales et humaines.* Québec: Les Publications du Québec, 2000.

Green, Lawrence, Ian Miles, and Jason Rutter. *Hidden Innovation in the Creative Industries.* NESTA Working Paper. London: NESTA, 2007.

Greenberg, Daniel S. *Science, Money, and Politics: Political Triumph and Ethical Erosion.* Chicago: University of Chicago Press, 2001.

Gurrutxaga, Ander. *Recorridos por el cambio, la innovación y la incertidumbre.* Bilbao: Universidad del País Vasco, 2009.

———. "Recorridos de la innovación." In *¿Cómo es una sociedad innovadora?.* Zamudio: Innobasque, 2009.

———. "Pathways to Social Innovation in the Basque Country." In *Implications of Current Research on Social Innovation in the Basque Country*, edited by Ander Gurrutxaga and Antonio Rivera. Reno: Center for Basque Studies, University of Nevada, Reno, 2011.

———. "Voces y argumentos de innovación social." In *La luz de la luciérnaga. Diálogos de innovación social*, edited by Ander Gurrutxaga and Javier Echeverría. Madrid: Plaza y Valdés, 2012.

Gurrutxaga, Ander, and Javier Echeverría. *La luz de la luciérnaga. Diálogos de innovación social.* Madrid: Plaza y Valdés, 2012.

Hacking, Ian. *Representing and Intervening: Introductory Topics in the Philosophy of Natural Science.* Cambridge: Cambridge University Press, 1983.

Hartmann, Stephan. "The World as Process: Simulations in the Natural and Social Sciences." In *Modelling and Simulation in the Social Sciences from the Philosophy of Science Point of View*, edited by Rainer Hegselmann, Ulrich Mueller, and

Klaus G. Troitzsch. Dordrecht: Kluwer, 1996.

Hevly, Bruce. "Reflections on Big Science and Big History." In *Big Science: The Growth of Large-Scale Research*, edited by Peter Galison and Bruce Hevly. Stanford: Stanford University Press, 1992.

Higgs, Peter L., and Stuart D. Cunningham. *Australia's Creative Economy: Mapping Methodologies*. Brisbane: ARC Centre of Excellence for Creative Industries & Innovation, 2007.

———. "Creative Industries Mapping: Where Have We Come From and Where are We Going?" *Creative Industries Journal* 1, no. 1 (2009): 7–30.

Higgs, Peter L, Stuart D. Cunningham, and Janet D. Pagan. *Australia's Creative Economy: Basic Evidence on Size, Growth, Income and Employment*. Technical Report. Brisbane: ARC Centre of Excellence for Creative Industries & Innovation, 2007.

Higgs, Peter L., Stuart D. Cunningham, and Hasan Bakshi. *Beyond the Creative Industries: Mapping the Creative Economy in the United Kingdom*. NESTA Technical Report BCI/07. London: NESTA, 2008.

HM Treasury, the DTI, and the Department for Education and Skills. *Science and Innovation Investment Framework 2004–2014*. London: The Stationery Office, 2004.

Hochgerner, Josef. "Innovation Processes in the Dynamics of Social Change." In *Innovation Cultures: Challenge and Learning Strategy*, edited by Jiři Loudin and Klaus Schuch. Prague: Filosofia, 2009.

Hoddeson, Lillian. "Mission Change in the Large Laboratory: The Los Alamos Implosion Program, 1943–1945." In *Big Science: The Growth of Large-Scale Research*, edited by Peter Galison and Bruce Hevly. Stanford: Stanford University Press, 1992.

Hogeschool vor de Kunsten Utrecht (HKU). *The Entrepreneurial Dimension of the Cultural and Creative Industries*. Utrecht, Netherlands: Hogeschool vor de Kunsten Utrecht, 2010.

Hollanders, Hugo, and Adriana van Cruysen. *Rethinking the European Innovation Scoreboard: A New Methodology for 2008–*

2010. PRO INNO Europe. N.p.: Innometrics, 2008.

Hollanders, Hugo, and Stefano Tarantola. *Innovation Union Scoreboard 2010: Methodology Report*. N.p.: Pro InnoEurope, 2011.

Hollanders, Hugo, Stefano Tarantola, and Alexander Loschky. *Regional Innovation Scoreboard 2009: Methodology Report*. N.p.: Innometrics: 2009.

Hottois, Gilbert. *La philosophie des technosciences*. Abidjan: Presses des Universités de Côte d'Ivoire, 1997.

Howaldt, Jürgen. "Shaping Social Innovation by Social Research." Paper presented to Challenge Social Innovation conference, Vienna, Austria, September 19–21, 2011. www.socialinnovation2011.eu/papers-abstracts.

Howaldt, Jürgen, and Michael Schwarz. *Social Innovation: Concepts, Research Fields and International Trends*. Dortmund, Germany: Sozialforschungsstelle, ZWE der TU-Dortmund, 2010. www.asprea.org/imagenes/IMO%20Trendstudie_Howaldt_englisch_Final%20ds.pdf.

Howkins, John. *The Creative Economy: How People Make Money From Ideas*. London: Penguin, 2001.

Hubert, A., ed. *Empowering People, Driving Change: Social Innovation in the European Union*. Brussels, BEPA 2013, 2010.

Hughes, Alastair, Kyla Moore, and Nimesh Kataria. *Innovation in Public Sector Organisations: A Pilot Survey for Measuring Innovation across the Public Sector*. London: NESTA, 2011.

Hui, Desmond. *Baseline Study of Hong Kong's Creative Industries*. Hong Kong: HKSAR Government Central Policy Unit, 2003.

Ilie, Elisabeta G., and Roel During. *An Analysis of Social Innovation Discourses in Europe: Concepts and Strategies of Social Innovation*. Wageningen, Netherlands: Alterra Institute, Wageningen UR, 2012.

Innenarity, Daniel, and Ander Gurrutxaga. *¿Cómo es una sociedad innovadora?* Zamudio, Basque Country, Spain: Innobasque, 2009.

Isaacson, Jeffrey A., Christopher Lane, and John Arquilla. *Predicting Military Innovation*. Santa Monica, CA: Rand Corporation, 1999.

Jacobs, Jane. *The Economy of Cities*. New York: Random House, 1969.

James, Saffron, and Nusrat Faizullah, with Nicola Bacon, Liz Bartlett, Neal Harrison, and Geoff Mulgan. *Making the Most of Local Innovation: What Makes Places Innovative and How Local Innovations Can Best Be Exploited*. Interim Report. London: Young Foundation and NESTA, 2007.

Jasanoff, Sheila. *Science at the Bar: Law, Science, and Technology in America*. Cambridge, MA: Harvard University Press, 1995.

Johannessen, Jon-Arild, Bjørn Olsen, and G.T. Lumpkin. "Innovation as Newness: What is New, How New, and New to Whom?" *European Journal of Innovation Management* 4, no. 1 (2001): 20–31.

Johnson, Steven. *Emergence: The Connected Lives of Ants, Brains, Cities, and Software*. New York: Free Press, 2001.

———. *Where Good Ideas Come From: The Natural History of Innovation*. New York: Riverhead Books, 2010.

KEA European Affairs. *The Economy of Culture in Europe*. Study prepared for the European Commission (Directorate-General for Education and Culture). With the support of the Turku School of Economic and Business Administration and MKW Wirtschaftsforschung GMbH. Brussels: KEA European Affairs, 2006.

———. "Towards a Culture-based Creativity Index." Paper presented at the Conference Measuring Creativity, Brussels, Belgium, May 28–29, 2009. www.keanet.eu/report/measuringcreativity.pdf.

———. *Business Innovation Support Services for Creative Industries*. Report for the European Commission (DG Enterprise and Industry). Brussels: KEA European Affairs, 2010.

Kern, Philippe. "The Economy of Culture – Proposal for a European Strategy." Transcript of speech given at the Cultural and Creative Industries in Europe: Coherent Policy in a Globalised World conference, Berlin, Germany, May 4, 2007. Brussels: KEA European Affairs, 2007. www.keanet.eu/report/KEA%20-%20Berlin%20presentation%20

May2007.pdf.

Klein, Juan-Luis, and Denis Harrison, eds. *L'innovation sociale. Émergence et effets sur la transformation des sociétés*. Québec: Presses de l'Université de Québec, 2006.

Kim, Linsu. *Imitation to Innovation: The Dynamics of Korea's Technological Learning*. Boston: Harvard Business School Press, 1997.

Kline, Stephen J., and Nathan Rosenberg. "An Overview of Innovation." In *The Positive Sum Strategy: Harnessing Technology for Economic Growth*, edited by Ralph Landau and Nathan Rosenberg. Washington, DC: National Academy Press, 1986.

KMU Forschung Austria (KMFA). "Creative Industries in Austria. Erster Oesterreicher Kreativwirtschaftsbericht." Vienna: KMU Forschung Austria (KMFA), 2003.

Kropotkin, Peter. *Mutual Aid, a Factor of Evolution*. Edited and with an introduction by Paul Avrich. New York: New York University Press, 1972 [1902].

Kuhn, Thomas S. *The Essential Tension: Selected Studies in Scientific Tradition and Change*. Chicago: University of Chicago Press, 1977. Latour, Bruno. *Science in Action: How to Follow Scientists and Engineers through Society*. Harvard: Harvard University Press, 1987 [1983].

———. *Reassembling the Social: An Introduction to Actor-Network-Theory*. Oxford: Oxford University Press, 2005.

Laudan, Larry. *Science and Values*. Berkeley: University of California Press, 1984.

Leibniz, G. (), ed. Gerhardt (GP), Philosophische Schriften. Hildesheim, Olms, 1960-61, 7 vols.

Liljemark, Thomas. *Innovation Policy in Canada: Strategy and Realities*. R2004: 019. Östersund: ITPS Swedish Institute For Growth Policy Studies, 2005.

Lundvall, Bengt-Åke. *Product Innovation and User-Producer Interaction*. Aalborg, Denmark: Aalborg University Press, 1985.

———, ed. *National Systems of Innovation: Towards a Theory of Innovation and Interactive Learning*. London: Pinter, 1992.

———. "Innovations as an Interactive Process: From User-Producer Interaction to the National System of Innovation." In *Technical Change and Economic Theory*, edited by Giovanni Dosi, Christopher Freeman, Richard Nelson, Gerald Silverberg, and Luc Soete. London: Pinter, 1998.

Maidagán, María Jesús, Iñaki Ceberio, Luis Garagalza, and Gotzon Arrizabalaga, eds. *Filosofía de la innovación. El papel de la creatividad en un mundo global.* Madrid: Plaza y Valdés, 2009.

Margulis, Lynn, and Dorion Sagan. *Microcosmos: Four Billion Years of Evolution from our Microbial Ancestors.* Foreword by Lewis Thomas. New York: Summit Books, 1986.

Martín Municio, Ángel, ed. *El valor económico de la lengua española.* Madrid: Espasa, 2003.

Martínez, Sergio, Xiang Huang, and Godfrey Guillamin, eds. *Historia, Prácticas y Estilos en la Filosofía de la Ciencia: Hacia una epistemología plural.* Mexico City: Porrúa, 2011.

McCartney, Scott. *ENIAC: The Triumphs and Tragedies of the World's First Computer.* Berkeley: Berkeley Publishing Group, 2001.

Miles, Ian, and Lawrence Green. *Hidden innovation in the Creative Industries.* Research Report. London: NESTA, 2008.

Ministry for Culture and Heritage/Te Manatū Taonga. *Cultural Indicators for New Zealand/Tohu Ahurea mō Aotearoa.* Wellington, New Zealand: Ministry for Culture and Heritage/Te Manatū Taonga, 2009.

Ministry of Industry, Employment and Communications and the Ministry of Education (Swedish Government). *Innovative Sweden: A Strategy for Growth through Renewal.* Ds2004 36. Stockholm: Ministry of Industry, Employment and Communications, 2004.

Mitcham, Carl. *Thinking through Technology: The Path Between Engineering and Philosophy.* Chicago: University of Chicago Press, 1994.

Mosterín, Jesús. *Conceptos y teorías en la ciencia.* Madrid: Alianza, 1984.

Moulaert, Frank, and Abdelillah Hamdouch. "New Views of Innovation Systems." *The European Journal of Social Science*

Research 19, no. 1 (2006): 11–24.

Moulines, C. Ulises. *Exploraciones metacientíficas. Estructura y contenido de la ciencia*. Prologue by Jesús Mosterín. Madrid: Alianza, 1982.

Moya, Eugenio. *Crítica de la razón tecnocientífica*. Madrid: Biblioteca Nueva, 1998.

Mulgan, Geoff. *Ready or Not: Taking Innovation in the Public Sector Seriously*. London: Young Foundation, 2007.

Mulgan, Geoff, with Simon Tucker, Rushanara Ali, and Ben Sanders. *Social Innovation: What Is It, Why It Matters, and How It Can Be Accelerated*. London: Young Foundation; Basingstoke Press, 2007.

Mulgan, Geoff, with Rushanara Ali, Richard Halkett, and Ben Sanders. *In and Out of Sync: The Challenge of Growing Social Innovations*. Research Report. London: NESTA, 2007.

Mulgan, Geoff, and David Albury. *Innovation in the Public Sector*. Discussion Paper. London: Strategy Unit, Cabinet Office, 2003.

Murray, Robin, Julie Caulier-Grice, and Geoff Mulgan. *The Open Book of Social Innovation*. London: Young Foundation and NESTA, 2010.

Nelson, Richard R., ed. *National Innovation Systems: A Comparative Analysis*. Oxford: Oxford University Press, 1993.

Nelson, Richard R., and Sidney G. Winter. "In Search of a Useful Theory of Innovation." *Research Policy* 6, no. 1 (1977): 36–76.

———. *An Evolutionary Theory of Economic Change*. Harvard, MA: Harvard University Press; London: Belknap Press, 1982.

NESTA. *Creating Growth: How the UK Can Develop World Class Creative Business*. NESTA Research Report. London, NESTA, 2006.

———. *The Innovation Gap: Why Policy Needs to Reflect the Reality of Innovation in the UK*. Policy Briefing IG/02, NESTA Policy and Research Unit. London: NESTA, 2006.

———. *Creative Entrepreneurship: Higher Education and the Creative Industries*. London: NESTA and the Higher Education Academy Art Design Media Subject Centre, 2007.

———. *Hidden Innovation: How Innovation Happens in Six 'Low Innovation' Sectors*. London: NESTA, 2007.

———. *Beyond the Creative Industries: Making Policy for the Creative Economy*. Policy Briefing BCI/20. London: NESTA, 2008.

———. *Creating Innovation: Do the Creative Industries Support Innovation in the Wider Economy?* London: NESTA, 2008.

———. *The Innovation Index: Measuring the UK's Investment in Innovation and its Effects*. Index Report. London: NESTA, 2009.

New Zealand Institute of Economic Research (NZIER). *The Creative Sector in New Zealand: Mapping and Economic Role*. A Report to New Zealand Trade & Enterprise. Wellington: NZIER, 2009.

Nonaka, Ikujiro, and Hirotaka Takeuchi. *The Knowledge-Creating Company: How Japanese Companies Create the Dynamics of Innovation*. Oxford: Oxford University Press, 1995.

Nowotny, Helga, Peter Scott, and Michael T. Gibbons. *Re-thinking Science: Knowledge and the Public in the Age of Uncertainty*. Cambridge: Polity, 2001.

Oakley, Kate, Brooke Sperry, and Andy C. Pratt. *The Art of Innovation: How Fine Arts Graduates Contribute to Innovation*. Edited by Hasan Bakshi. NESTA Research Report 16. London: NESTA, 2008.

OECD (Organisation for Economic Co-operation and Development). *OECD Proposed Guidelines for Collecting and Interpreting Technological Innovation Data: Oslo Manual*. Paris: OECD, 1992.

———. *National Innovation Systems*. Paris: OECD, 1997.

———. *Science, Technology and Industry Review*. Special Issue on New Science and Technology Indicators, No. 27. Paris: OECD, 2001.

———. *Dynamising National Innovation Systems*. Paris: OECD, 2002.

———. *Frascati Manual: Proposed Standard Practice for Surveys on Research and Experimental Development*. Paris: OECD, 2002.

———. *Science, Technology and Innovation Indicators for a Changing World: Responding to Policy Needs*. Paris: OECD, 2007.
———. *The New Nature of Innovation*. Paris: OECD, 2009.
———. *OECD Patent Manual*. Paris: OECD, 2009.
———. *Science, Technology and Industry Scoreboard 2009*. Paris: OECD, 2009.
———. *Main Science and Technology Indicators*. Volume 2010, Issue 1. Paris: OECD, 2010.
———. *Measuring Innovation: A New Perspective*. Paris: OECD, 2010.
———. *OECD Science, Technology and Industry Outlook 2010*. Paris: OECD, 2010.
———. *SMEs, Entrepreneurship and Innovation*. Paris: OECD, 2010.
———. *The OECD Innovation Strategy: Getting a Head Start on Tomorrow*. Paris: OECD, 2010.
OECD and Eurostat. *The Measurement of Human Resources Devoted to Science and Technology: Canberra Manual*. Paris: OECD, 1995.
———. *Proposed Guidelines for Collecting and Interpreting Technological Innovation Data: Oslo Manual*. 2nd ed. Paris: OECD/EC, 1997.
———. *Oslo Manual: Guidelines for Collecting and Interpreting Innovation Data*. 3rd ed. Paris: OECD, 2005.
OECD and FORA. *New Nature of Innovation*. Copenhagen: FORA; OECD, 2009.
Open Method of Coordination-Expert Working Group (OMC-EWG). *Final report: OMC – Expert Working*
Group on Maximising the Potential of Cultural and Creative Industries, in Particular that of SMEs. Brussels: OMC-EWG, 2010.
Panofsky, W. K. H. "SLAC and Big Science: Stanford University." In *Big Science: The Growth of Large-Scale Research*, edited by Peter Galison and Bruce Hevly. Stanford: Stanford University Press, 1992.
Petitclerc, Martin. *Rapport sur les innovations sociales et les transformations sociales*. Les cahiers du Crises. Collection Études théoriques, ET0313. Québec: Crises, 2003.

Phills Jr., James. A., Kris Deiglmeier, and Dale T. Miller. "Rediscovering Social Innovation." *Stanford Social Innovation Review*, online journal (Fall 2008): 34–43.

Pickering, Andrew, ed. *Science as Practice and Culture*. Chicago: University of Chicago Press, 1992.

Pickering, Andrew. *The Mangle of Practice: Time, Agency, and Science*. Chicago: University of Chicago Press, 1995.

Polanyi, Michael. *The Tacit Dimension*. London: Routledge, 1966.

Porter, Michael E. *Competitive Advantage: Creating and Sustaining Superior Performance*. New York: Free Press, 1985.

———. *The Competitive Advantage of Nations*. New York: Free Press, 1990.

Potts, Jason. "Introduction: Creative Industries & Innovation Policy." *Innovation: Management, Policy & Practice* 11, no. 2 (2009): 138–47.

Potts, Jason, and Stuart D. Cunningham. "Four Models of the Creative Industries." *International Journal of Cultural Policy* 14, no. 3 (2008): 233–47.

Potts, Jason, Stuart D. Cunningham, John Hartley, and Paul Ormerod. "Social Network Markets: A New Definition of Creative Industries." *Journal of Cultural Economics* 32, no. 3 (2008): 167–85.

Power, Dominic, and Tobias Nielsén. *Priority Sector Report: Creative and Cultural Industries*. European Commission, Enterprise and Industry, Europe Innova, and the European Cluster Observatory. Stockholm: European Cluster Observatory, 2010.

Power, D., and T. Nielsén (2010b), Priority Sector Report: Creative and Cultural Industries – Methodological Appendix-, Estocolmo, European Cluster Observatory (ECO), ProInnova.

Pratt, Andy C. *The Cultural Industries Sector: Its Definition and Character from Secondary Sources on Employment and Trade, Britain 1984–91*. Online Research Paper. Research Papers in Environmental and Spatial Analysis 41, Department of Geography, London School of Economics and Political Science, 1997.

———. "Cultural Industries and Public Policy." *International Journal of Cultural Policy* 11, no. 1 (2005): 31–44.

———. "Creative Cities: The Cultural Industries and the Creative Class." *Geografiska Annaler*. Series B, *Human Geography* 90, no. 2 (2008): 107–17.

Preskill, Hallie, and Tanya Beer. *Evaluating Social Innovation*. Washington, DC: FSG and the Center for Evaluation Innovation, 2012.

Putnam, Hilary. *The Collapse of the Fact/Value Dichotomy*. Cambridge: Harvard University Press, 2004.

Rapp, Friedrich. *Analytical Philosophy of Technology*. Translated by Stanley R. Carpenter and Theodor Langenbruch. Dordrecht, Netherlands: D. Reidel Pub. Co., 1981.

Ravetz, Jerome R. *Scientific Knowledge and Its Social Problems*. Oxford: Clarendon Press, 1971.

Reid, Benjamin, Alexandra Albert, and Laurence Hopkins. *A Creative Block? The Future of the UK Creative Industries*. A Knowledge Economy & Creative Industries Report. London: The Work Foundation, 2010.

Reinert, Hugo, and Erik S. Reinert. "Creative Destruction in Economics: Nietzsche, Sombart, Schumpeter." In *Friedrich Nietzsche (1844–1900): Economy and Society*, edited by Jürgen G. Backhaus and Wolfgang Dreschler. New York: Springer, 2006.

Rescher, Nicholas. "On Situating Process Philosophy." *Process Studies* 28, nos. 1–2 (1999): 37–42.

———. "The Promise of Process Philosophy." In *The Edinburgh Companion to Twentieth-Century Philosophies*, edited by Constantin V. Boundas. Edinburgh: Edinburgh University Press, 2007.

RICYT. *Estado de Ciencia: Principales Indicadores de Ciencia y Tecnología - Iberoamericanos / Interamericanos*. Buenos Aires: RICYT, 2010. www.ricyt.org/index.php?option=com_content&view=article&id=211:el-estado-dela-ciencia-2010&catid=6:publicaciones&Itemid=7.

RICYT/OEA/CYTED. *Standardization of Indicators of Technological Innovation in Latin American and Caribbean Countries:*

Bogotá Manual. Buenos Aires: RICYT, 2001.

Rodríguez Herrera, Adolfo, and Hernán Alvarado Ugarte. *Claves de la innovación social en América Latina y el Caribe*. Santiago de Chile: Comisión Económica para América Latina y el Caribe (CEPAL), 2008.

Rogers, Everett M. *Diffusion of Innovations*, 4th ed. New York: The Free Press, 1995.

Rouse, Joseph. *Engaging Science: How to Understand Its Practices Philosophically*. Ithaca, NY: Cornell University Press, 1996.

Sánchez Ron, José Manuel. *Falsos mitos: ciencia versus tecnología: Reflexiones sobre política científica*. Madrid: Fundación Repsol, 1998.

Sainsbury, Lord James. Science & Innovation Investment Framework 2004-2014. London: HM Treasury & Department for Education and Skills, 2004.

Scott, Allen J. *The Cultural Economy of Cities: Essays on the Geography of Image-Producing Industries*. London: SAGE, 2000.

Schumpeter, Joseph A. *The Theory of Economic Development: An Inquiry into Profits, Capital, Credit, Interest, and the Business Cycle*. Translated from German by Redvers Opie. Boston: Harvard University Press, 1934.

———. *Business Cycles*. New York: McGraw-Hill, 1939. Online edition by Didier Lagrange, L'université du Québec à Chicoutimi (UQAC), Québec, Canada, 2008 (accessed August 15, 2011).

———. *Capitalism, Socialism and Democracy*. New York: Harper & Row, 1975 [1942].

———. "The Creative Response in the Economic History." *Journal of Economic History* 7, no. 2 (November 1947): 149–59.

———. "Economic Theory and Entrepreneurial History." In *Change and the Entrepreneur: Postulates and the Patterns for Entrepreneurial History*, edited by Harvard University Research Center in Entrepreneurial History. Cambridge, MA: Harvard University Press, 1949.

———. *Essays*. Cambridge: Addison-Wesley Press, 1951.

Sebastián, Jesús. "¿Hay alguna innovación que no sea social?" In *Contextos y usos de la innovación social*, edited by Lucía Merino. Leioa: Servicio Editorial de la Universidad del País Vasco, 2012.

Seidel, R. "The Origins of the Lawrence Berkeley Laboratory." In *Big Science: The Growth of Large-Scale Research*, edited by Peter Galison and Bruce Hevly. Stanford: Stanford University Press, 1992.

Sen, Amartya. *Development as Freedom*. Oxford: Oxford University Press, 1999.

Simon, Herbert A. *The Sciences of the Artificial*. Cambridge, MA: MIT Press, 1996.

Social Innovation eXchange (SIX) and the Young Foundation. *Study on Social Innovation*. N.p.: European Union/Young Foundation, 2010.

Smith, Bruce L.R. *American Science Policy since World War II*. Washington, DC: The Brookings Institution, 1990.

Smith, Keith. "Measuring Innovation." In *The Oxford Handbook of Innovation*, edited by Jan Fagerberg, David C. Mowery, and Richard R. Nelson. Oxford: Oxford University Press, 2005.

Smith, Robert W. "The Biggest Kind of Big Science: Astronomers and The Space Telescope." In *Big Science: The Growth of Large-Scale Research*, edited by Peter Galison and Bruce Hevly. Stanford: Stanford University Press, 1992.

"Social Innovation in a Post-Crisis World." Special edition for the World Economic Forum Annual Meeting. *Innovations: Technology, Governance, Globalization*. MIT Press Journal. Davos-Klosters, 2009.

Solla Price, Derek J. de. *Little Science, Big Science*. New York: Columbia University Press, 1962.

Stoneman, Paul. *Soft Innovation: Towards a More Complete Picture of Innovative Change*. NESTA Research Report 29. London: NESTA, 2009.

Taipale, Ilkka, ed. *100 sosiaalista innovaatiota Suomesta*. Helsinki: Itämerikeskussäätiö: Kunnia, 2006.

Tarde, Gabriel. *The Laws of Imitation*. Translated by Elsie Clew Parsons. With an introduction by Franklin H. Giddings. New York: Henry Holt & Co., 1903.

Taylor, James B. "Introducing the Social Innovation." *Journal of Applied Behavioral Science* 6, no. 1 (1970): 69–77.

TERA Consultants. *Building a Digital Economy: The Importance of Saving Jobs in the EU's Creative Industries*. Paris: TERA Consultants, 2010.

Throsby, David. *Economics and Culture*. Cambridge: Cambridge University Press, 2001.

Towse, Ruth, ed. *A Handbook of Cultural Economics*. Cheltenham, UK: Edward Elgar, 2003.

UNCTAD (United Nations Conference on Trade and Development). *Creative Industries and Development*. Document TD(XI)/BP/13. Geneva: United Nations, 2004.

———. *Creative Economy Report 2008: The Challenge of Assessing the Creative Economy: Towards Informed Policy-making*. UNCTAD/DITC/2008/2. Geneva: United Nations, 2008.

———. *Creative Economy Report 2010: Creative Economy: A Feasible Development Option*. Geneva: United Nations, 2010.

UNESCO (United Nations Educational, Scientific and Cultural Organization). *Cultural Industries*. Paris: UNESCO, 1982.

———. *Framework for Cultural Statistics*. Paris: UNESCO, 1986. www.uis.unesco.org/Culture/Pages/framework-cultural-statistics.aspx .

———. *2009 UNESCO Framework for Cultural Statistics*. Montreal: UNESCO Institute for Statistics (UIS), 2009.

UNIDO (United Nations Industrial Development Organization). *Creative Industries and Micro & Small Scale Enterprise Development: A Contribution to Poverty Alleviation*. XP/RAS/05/002. A Joint Initiative by UNIDO and UNESCO. Vienna: UNIDO, 2011.

Von Hippel, Eric. "Lead Users: A Source of Novel Product Concepts." *Management Science* 32, no. 7 (1986): 791–805.

———. *The Sources of Innovation*. New York: Oxford University Press, 1988.

———. *Democratizing Innovation*. Cambridge, MA: MIT Press, 2005.

Waal, Frans de. *Good Natured: The Origins of Right and Wrong in*

Humans and Other Animals. Cambridge, MA: Harvard University Press, 1996.
Weinberg, Alvin M. "Impact of Large-scale Science on the United States." *Science* 134 (1961): 161–64.
———. *Reflections on Big Science*. Cambridge, MA: MIT Press, 1967.
Whitehead, Alfred North. *The Concept of Nature*. Cambridge: Cambridge University Press, 1920.
Wilkinson, Alex. *An Assessment of Productivity Indicators for the Creative Industries*. London: Department for Culture, Media and Sport, 2007.
Wolfram, Stephen. *A New Kind of Science*. Urbana-Champaign, IL: Wolfram Media Inc., 2002.
Work Foundation, The. *Staying Ahead: The Economic Performance of the UK's Creative Industries*. London: The Work Foundation, 2007.
World Intellectual Property Organisation (WIPO). *Guide on Surveying the Economic Contribution of the Copyright-Based Industries*. Geneva: World Intellectual Property Organisation, 2003.
Ziman, J.M. *Real Science: What It Is, and What It Means*. Cambridge: Cambridge University Press, 2000.
ZSI (Zentrum für Soziale Innovation).
"The Analysis of Social Innovations as Social Practice." Research paper prepared by Josef Hochgerner. Vienna: ZSI Centre for Social Innovation, 2011.

INDEX

Page numbers followed by *f* indicate figures. Numbers followed by n indicate footnotes.

Made in United States
North Haven, CT
10 November 2026

22091345R00229